Springer Theses

Recognizing Outstanding Ph.D. Research

Aims and Scope

The series "Springer Theses" brings together a selection of the very best Ph.D. theses from around the world and across the physical sciences. Nominated and endorsed by two recognized specialists, each published volume has been selected for its scientific excellence and the high impact of its contents for the pertinent field of research. For greater accessibility to non-specialists, the published versions include an extended introduction, as well as a foreword by the student's supervisor explaining the special relevance of the work for the field. As a whole, the series will provide a valuable resource both for newcomers to the research fields described, and for other scientists seeking detailed background information on special questions. Finally, it provides an accredited documentation of the valuable contributions made by today's younger generation of scientists.

Theses are accepted into the series by invited nomination only and must fulfill all of the following criteria

- They must be written in good English.
- The topic should fall within the confines of Chemistry, Physics, Earth Sciences, Engineering and related interdisciplinary fields such as Materials, Nanoscience, Chemical Engineering, Complex Systems and Biophysics.
- The work reported in the thesis must represent a significant scientific advance.
- If the thesis includes previously published material, permission to reproduce this must be gained from the respective copyright holder.
- They must have been examined and passed during the 12 months prior to nomination.
- Each thesis should include a foreword by the supervisor outlining the significance of its content.
- The theses should have a clearly defined structure including an introduction accessible to scientists not expert in that particular field.

More information about this series at http://www.springer.com/series/8790

Álvaro González García

Polymer-Mediated Phase Stability of Colloids

Doctoral Thesis accepted by
Utrecht University, Utrecht, The Netherlands

Author
Dr. Álvaro González García
Van 't Hoff Laboratory for Physical and Colloid Chemistry, Department of Chemistry and Debye Institute
Utrecht University
Utrecht, The Netherlands

Supervisor
Prof. Remco Tuinier
Laboratory of Physical Chemistry, Department of Chemical Engineering and Chemistry and Institute for Complex Molecular Systems (ICMS)
Eindhoven University of Technology
Eindhoven, The Netherlands

ISSN 2190-5053 ISSN 2190-5061 (electronic)
Springer Theses
ISBN 978-3-030-33685-1 ISBN 978-3-030-33683-7 (eBook)
https://doi.org/10.1007/978-3-030-33683-7

© Springer Nature Switzerland AG 2019
This work is subject to copyright. All rights are reserved by the Publisher, whether the whole or part of the material is concerned, specifically the rights of translation, reprinting, reuse of illustrations, recitation, broadcasting, reproduction on microfilms or in any other physical way, and transmission or information storage and retrieval, electronic adaptation, computer software, or by similar or dissimilar methodology now known or hereafter developed.
The use of general descriptive names, registered names, trademarks, service marks, etc. in this publication does not imply, even in the absence of a specific statement, that such names are exempt from the relevant protective laws and regulations and therefore free for general use.
The publisher, the authors and the editors are safe to assume that the advice and information in this book are believed to be true and accurate at the date of publication. Neither the publisher nor the authors or the editors give a warranty, expressed or implied, with respect to the material contained herein or for any errors or omissions that may have been made. The publisher remains neutral with regard to jurisdictional claims in published maps and institutional affiliations.

This Springer imprint is published by the registered company Springer Nature Switzerland AG
The registered company address is: Gewerbestrasse 11, 6330 Cham, Switzerland

Supervisor's Foreword

This year it is exactly 65 years ago that Sho Asakura and Fumio Oosawa in Nagoya, Japan, showed that adding nonadsorbing macromolecules to a colloidal dispersion induced effective attractions between the colloidal particles. In these 65 years, this so-called depletion interaction has proven to be of great significance for colloidal phase stability.

The Ph.D. thesis of Álvaro González García describes the work he performed the past few years at Utrecht University and Eindhoven University of Technology, the Netherlands, which can partly be seen as an extension of the insights of the depletion interaction and resulting phase stability.

Álvaro graduated from his M.Sc. study with honours at Utrecht University, always focused on theory and simulation work of soft matter, with an internship at DSM where he also performed experiments. He demonstrated his ability to quick thinking and learning, by rapidly digging into and mastering the necessary thermodynamics of colloids to be able to do computations and, from the beginning, started to add his own ideas to the problems we worked on. His contributions to the field include the influence of direct interactions (beyond hardcore) between colloids in colloid–polymer mixtures, accurate estimations of free volume available in dense suspensions, the transition between polymer depletion and adsorption in colloid–polymer mixtures, polymer-mediated interactions between block copolymer micelles, and last but not least, extended insights into the phase behaviour of disc–polymer and superball–polymer mixtures. It was clear during the project that Álvaro had a keen interest in anisotropic particles. For that reason, he started to work on mixtures of platelets plus depletants (Chap. 6). At some point, he entered my office with the question: is it possible to have four-phase coexistence in platelet–polymer mixtures? This question made me think. That unique moment that one finds something unexpected…

Several topics he worked on are topics I worked on and thought about for maybe a decade or so (Chaps. 2–5, 8), but which were solved by him to a, in my opinion, satisfactory degree. Very rapidly within the Ph.D. project of Álvaro our working relationship was developing from supervisor-student towards a collaborator role.

He was always up for any new challenge I gave him but also started to have more own ideas.

I am grateful and pleased Springer is publishing this exceptional thesis of Álvaro González García and thank Eindhoven University of Technology and Utrecht University for encouraging interactions between our universities, which stimulated this work.

Eindhoven, The Netherlands
July 2019

Prof. Remco Tuinier

Preface

I said to the almond tree, "Sister, speak to me of God".
And the almond tree blossomed.
Nikos Kazantzakis

One may pretend to understand the complex processes that ultimately give the almond tree flower its functional aesthetic characteristics. Such an act of vanity reflects the opposite of the scientific method. In order to grasp the real world, one must simplify it. An effective approach is to select a small number of relevant parameters, and test how robust the predictions extracted from the experimental, computational or theoretical method followed are compared with real-life observations. Note that even experimental laboratory research provides a simplified picture of real systems.

In this thesis, we have isolated some key parameters governing the (in)stability of colloid–polymer mixtures. The term 'colloid' refers to a state of matter in which a certain amount of material (with one of its dimensions between one nanometre to one micrometre) is dispersed in another medium. Polymers are macromolecules constituted of many repeating units called segments; depending on the number and nature of these segments, polymers may dissolve or phase-separate in solution. Colloid–polymer mixtures are widespread in biological systems (including blood, the cytoplasm of a living cell, and plant sap), as well as in man-made products (such as paints, drinking yogurt, and printing inks). Better control over the stability limits during product development is possible via a fundamental understanding of the effect of some relevant parameters in the system at hand. In the examples given above, multiple colloids, polymers, and other components are often present. Building knowledge on the interactions between components of the same nature, and pairs of different components is a logical starting point. Based on the characteristics of the colloidal particles investigated, we sequester this thesis into three parts.

In Part I, we took the simplest model system: mixtures of hard *spheres* (like billiard balls) with added polymers simplified as ghost-like spheres. We studied how a direct soft interaction beyond the hardcore interaction modulates the phase stability of a model colloid–polymer mixture (Chap. 2). We conclude that soft repulsive interactions widen the stable region and direct soft attractions decrease the stability of a colloid–polymer mixture. We also paid attention to mixtures of such colloidal hard spheres with added tiny polymers (Chap. 3): such a model system may be of relevance, for instance, in protein crystallisation. Upon revisiting a well-established (relatively simple) free volume theory, we improved it for the solid phase state, which brought it closer to more convoluted ones, simulations, and experiments. In Chaps. 2 and 3, we (over)simplified the polymers (we took them as a ghost-like sphere), while in Chap. 4 we describe them in more detail. In that chapter, we extract how the strength of the interaction between the surface of the colloidal particle and the polymer segments affects the phase behaviour of a colloid–polymer mixture. We elucidate the possibility of colloid–polymer mixtures which do not phase-separate, even at high polymer concentrations, which is appealing for industrial applications such as paint or foodstuff.

In Part II, we focus on the influence of the shape of the colloidal particle on the phase behaviour of colloid–polymer mixtures. Liquid crystalline phases in non-dilute colloidal dispersions may emerge as a consequence of *the anisotropic shape of particles*. Not surprisingly, the pigment's shape affects the final properties of paints and coatings. We consider anisotropic hard particles, and study the effects of adding ghost-like spheres to mimic polymer chains. Investigations of cube-like (Chap. 5) and platelet-like (Chap. 6) colloids reveal a rather rich phase behaviour. We highlight the unexpected presence of up to four phases in coexistence in effective two-component systems, reported for the first time in this thesis. Furthermore, we elucidate the relevance of compartmentalisation of tiny compounds in highly concentrated systems. This could be of interest, for instance, in the future development of photonic materials with two different optical paths, and may serve as a model to study crowded living environments.

Finally, in Part III, we studied *association colloids*. We focus on associative colloidal particles formed by diblock copolymers: polymers composed of well-soluble segments and of poorly soluble segments divided into two blocks. In a selective solvent, diblock copolymers can constitute the building blocks of equilibrium structures known as micelles. We focus on self-organised spherical micelles, used in applications ranging from cosmetics to targeted drug delivery to, for instance, tumoral cells. We studied micelle–micelle interactions, and particularly focus on how the building block composition affects colloidal stability (Chap. 7) and how it is affected by the addition of a second (non-blocky) polymer (Chap. 8). The associative and soft nature of association colloids render the problem at hand complex, yet insights could be extracted about the phase stability of micelles. We concluded that spherical micelles resulting from diblocks with a short soluble block are more suitable for applications than those with a large soluble block.

By virtue of these simplified models, a collection of predictions governing the (in)stability of colloid–polymer mixtures has been extracted. These may serve for further developments, considering, for instance, not only two but multi-component mixtures. Further tuning of the accuracy of the models could bring them closer to reality. The author hopes that the small pieces that this thesis has added to the puzzle of knowledge may inspire and be of utility to others.[1]

Utrecht, The Netherlands Álvaro González García

[1]Note: The first and last parts of this Summary are (heavily) influenced by the author's article in Cultural Resuena, 'Nikos Kazantzakis y el espíritu científico' (only in Spanish): http://www.culturalresuena.es/2016/10/kazantzakis-espiritu-cientifico/

Publications related to this thesis

This thesis is based upon the following publications:

- **Á. González García**, and R. Tuinier. Tuning the phase diagram of colloid–polymer mixtures via Yukawa interactions, *Phys. Rev. E*, **94**, 06260 (Chap. 2)
- **Á. González García**, J. Opdam, R. Tuinier, and M. Vis. Isostructural solid–solid coexistence of colloid–polymer mixtures, *Chem. Phys. Lett.*, **709**, 16–20 (Chap. 3)
- **Á. González García***, J. Nagelkerke*, R. Tuinier*, and M. Vis*. Unipletion in colloid–polymer mixtures, *in preparation* (Chap. 4)
- **Á. González García**, J. Opdam, and R. Tuinier. Phase behaviour of colloidal superballs mixed with non-adsorbing polymers, *Eur. Phys. J. E*; 41:110, 2018 (Chap. 5)
- **Á. González García**, H. H. Wensink, H. N. W. Lekkerkerker, and R. Tuinier. Entropic patchiness drives multi-phase coexistence in discotic colloid-depletant mixtures, *Sci. Rep.*, **7**, 17058, 2017 (Chap. 6)
- **Á. González García**, R. Tuinier, J. V. Maring, J. Opdam, H. H. Wensink, and H. N. W. Lekkerkerker. Depletion-driven four-phase coexistences in discotic systems, *Mol. Phys.*, **116**:21–22, 2757–2572, 2018 (Chap. 6)
- **Á. González García**, R. Tuinier, and A. Cuetos. Compartmentalisation in crowded discotics: quantifying what goes where, *in preparation* (Chap. 6)
- **Á. González García***, A. Ianiro*, and R. Tuinier. On the Colloidal Stability of Spherical Copolymeric Micelles, *ACS Omega* **3** (12), 17976–17985, 2018 (Chap. 7)
- **Á. González García***, A. Ianiro*, R. Beljon, Frans A. M. Leermakers, and R. Tuinier. Polymer-mediated stability of micellar suspensions, *in preparation* (Chap. 8)

Acknowledgements

This Ph.D. thesis would have not been possible without the guidance of my promotors, Profs. R. Tuinier and A. P. Philipse. I would also like to express my gratitude to the closest collaborators in the contents of this thesis: Alessandro Ianiro, Joeri Opdam, and Mark Vis. I would like to thank all my friends and colleagues at the Eindhoven University of Technology and Utrecht University who contributed to my Ph.D. experience beyond the merely professional. Finally, I thank my family for their constant support, and Sofia for being the equilibrium in my life.

Contents

Symbols and Acronyms

Symbols[2]

$\beta \equiv 1/k_B T$	The inverse of the thermal unit of energy $k_B T$, with k_B Boltzmann's constant and T the absolute temperature
r	Centre-to-centre distance between colloids
σ	Colloidal diameter ($\sigma \equiv 2R$, with R the colloidal radius)
W	Interaction potential
q_Y	Range of the HCY interaction
ϵ	Strength (contact potential) of the HCY interaction
$\hat{\kappa}$	Screening length of the HCY interaction
ϕ_d^R	Polymer bulk volume fraction
q	Relative size of polymer to colloidal particle
δ	Adsorption (and depletion) thickness
ζ	Strength of the penetrable sphere (PS) interaction
B_2	Second osmotic viral coefficient
v_c	Colloidal particle volume
$B_2^* \equiv B_2/v_c$	Normalised second virial coefficient
F	Helmholtz (free) energy
V	System volume
Λ_B	De Broglie thermal wavelength
ϕ_c	Colloid volume fraction
ϕ_c^{cp}	Colloid volume fraction at close packing
N_c	Number of colloids
$\gamma_Y, \gamma_1, \gamma_2, Q_Y, L_Y$	Set of equations defining the free energy following the FMSA
μ_i	Chemical potential of component i
Π	Osmotic pressure
Ω	(semi) grand-potential
α	Free volume fraction for depletants in the FVT framework

[2]Many symbols used in Sects. 3.2, Appendices 6.1 and 6.2 are intentionally left out of this list.

ϕ_d^S	Depletant volume fraction in the colloid-polymer mixture
v_d	Volume of a depletant
ω	Work
$\langle V_{free} \rangle_o$	Undistorted free volume for depletants in colloidal system
v_{exc}	Excluded volume
Q_s	Shape-dependent term in FVT upon applying SPT to the probability of inserting a depletant
$y \equiv \phi_c/(1-\phi_c)$	An auxiliary function commonly used in FVT derivations
N_d^R	Number of depletants in R considered in FVT
λ	Scaling factor of the SPT approach used to calculate α
κ	Total number of depletion zones overlaps within a UC
ϕ_c^*	ϕ_c from which depletion zones overlap for HSs in a FCC lattice
o	Undistorted properties (depletant-free system, isolated micelle)
ϕ_k	Volume fraction of components in the SCF-lattice, such that: ϕ_A, solvophilic block; ϕ_B, solvophilic block; ϕ_p, micelle forming polymer; ϕ_G, added guest homopolymer to the colloidal suspension; and ϕ_W, solvent (never graphically presented).
N_{lat}	Number of lattice sites considered (i.e., the size of the lattice)
b	Size of a lattice site
z	Lattice layer considered, independently of the lattice geometry
K	Number of nearest micelles from a central one
n	Solvophilic block length
m	Solvophobic block length
χ_{ij}	Flory–Huggins interaction parameter between segments i and j, with $\{i, j\} = \{A, B, G, W, C\}$, where C is a lattice site belonging to the colloidal particle
R_h	Hydrodynamic radius of a micelle
$\bar{\alpha}, \bar{\rho}$	Auxiliary functions used for evaluating R_h
ϕ_p^{bulk}	Micelle-forming polymer bulk concentration
ϕ_G^{bulk}	Guest (homo)polymer bulk concentration
ϕ_G^*	Guest (homo)polymer overlap concentration
g_p	Aggregation number of amphiphilic molecules in a micelle; the number of building blocks (unimers) per micelle
R_g	Radius of gyration of a homopolymer in solution
v_G	Volume of the guest homopolymer added to a colloidal suspension
h	Distance between two flat surfaces
$\Delta\chi$	Effective affinity surface–polymer
m	Shape parameter of a superball, $m = 2$ corresponds to a sphere and $m = \infty$ to a cube

$f(m)$ Function defining the volume of a superball
$\mathfrak{r}$ Maximum distance from the centre of a superball
s_{c} Surface area of a colloid
c_{c} Surface-integrated mean curvature of a colloid
$\mathfrak{r}$ Maximum distance from the centre of a superball
γ Asphericity parameter
$\mathfrak{Q}, \mathfrak{R}, \mathfrak{S}$ Functions of γ used for the EOS of *convex* particles
V_{f} Volume that a colloid explores within the UC without overlapping with others
V_{UC} Volume of the crystalline unit cell
$\Lambda \equiv L/\sigma$ Aspect ratio of a cylinder, with L its length
ρ_{c} Number density of colloid
$\rho_{\mathrm{d}}^{\mathrm{R}}$ Number density of depletants in bulk
$\tilde{z}$ Depletant fugacity
$\mathfrak{C}$ Number of components in a system
G_{P} Parsons–Lee scaling factor
$\|$ Parallel to the columnar direction vector (i.e., intra-columnar direction)
$\perp$ Perpendicular to the columnar direction vector (i.e., inter-columnar direction)
$r_{\|}$ Intra-columnar direction
$r_{\perp}$ Inter-columnar direction
$g_{\perp}^{i-j}$ Distribution function of particle-pair $\{i,j\}$ in $r_{\perp}$, with $\{i,j\}$ the colloid (c) or depletant (d)
$g_{\|}^{\mathrm{c-c},00}$ Colloid–colloid distribution function in $r_{\|}$ for colloids within the same column
$g_{\|}^{\mathrm{c-c},01}$ Colloid–colloid distribution function in $r_{\|}$ for colloids in different columns
$\Delta_{\perp}$ Spacing between platelets in $r_{\perp}$
$\Delta_{\|}$ Spacing between platelets in $r_{\|}$
$\phi_{\mathrm{c}}^{\|}$ ϕ_{c} from which depletion zones start to overlap in $r_{\|}$
$\phi_{\mathrm{c}}^{\perp}$ ϕ_{c} from which depletion zones start to overlap in $r_{\perp}$

Acronyms

ACPM Association colloid–polymer mixture
AOV Asakura–Oosawa–Vrij
C Columnar phase
CEP Critical end point
CMC Critical micelle concentration
cp Close packing
CP Critical point

CPM	Colloid–polymer mixture
CS	Carnahan–Starling EOS
EOS	Equation of state
F	Colloidal fluid phase
FCC	Face-entred cubic crystalline phase
FMSA	First-order mean spherical approximation
FVT	Free volume theory
G	Colloidal gas phase
HS	Hard sphere
I	Isotropic phase
L	Colloidal liquid phase
LJD	Lennard-Jones-Devonshire EOS
N	Nematic phase
ODF	Orientation distribution function
PHS	Penetrable hard sphere
QP	Quadruple point
R	FVT reservoir of depletants
S	System of interest
SC	Simple cubic crystalline phase
SCF	Self-consistent field, used here to reefer to the Scheutjens–Fleer SCF theory
SPT	Scaled particle theory
TP	Triple point
UC	Unit cell of a crystalline structure
VL	Vliegenthart–Lekkerkerker criterion

Chapter 1
Introduction

> All models are wrong, but some are useful.
> *George E. P. Box*

1.1 Colloids: Interactions and Phase Behaviour

Colloids are dispersions of a phase (solid, gas, or liquid) in another medium. Characteristic of colloidal particles is their size; in at least one dimension their length scale is between, say, a nm and a μm. One may also define the upper-size limit of colloidal particles by the fact that they exhibit Brownian motion [1]. If the thermal energy, which drives Brownian or thermal motion, is large compared to the gravitational energy, an ensemble of colloidal particles may reach equilibrium states similar to those manifested in atomic systems [2]. Presumably, at low enough densities, a colloidal dispersion behaves as an ideal gas; an ensemble of molecules at very low densities. This colloid–atom analogy was used by Einstein in his theory for Brownian motion of particles suspended in a solvent [3]. The later experimental verification by Perrin [4] of the colloidal barometric height distribution following Boltzmann's law constituted the starting point of systematic studies on the dynamics and equilibrium properties of colloidal systems [5].

In contrast to interactions between molecules or atoms, colloid–colloid interactions are tuneable, to a certain degree [6–9]. Control over the interactions may be achieved through modifications of the colloidal surfaces, such as functionalisation of the particles with polymers, surfactants or charged groups [10, 11]. A systematic method to effectuate well-defined attractions is through the addition of non-adsorbing polymers to a colloidal suspension [12–15]. In this case the polymers are excluded from a region near the particle surface, the so-called depletion zone, due to a loss of configurational entropy. In fact, the excluded volume between the polymer chain and the colloidal particle of interest defines this depletion zone. When two colloidal particles are brought in close proximity such that their depletion zones overlap, the

© Springer Nature Switzerland AG 2019

Á. González García, *Polymer-Mediated Phase Stability of Colloids*,
Springer Theses, https://doi.org/10.1007/978-3-030-33683-7_1

volume available for non-adsorbing polymers increases. This leads to an increase in entropy of the polymers. In general, a non-adsorbing species added to a colloidal suspension is termed 'depletant'. The magnitude and range of the resulting depletion attraction between particles are set by the concentration and size of the non-adsorbing depletant [14].

Colloidal particles can sometimes be envisaged as impenetrable (marble-like) spheres. The latter is known as the hard sphere (HS) model [16]. Silica or PMMA spheres in a refractive index-matched solvent are examples of experimental realisations of such a theoretically appealing model [17]. The hard-core repulsion between HSs is sufficient to explain the equilibrium fluid–solid phase transition solely on entropic grounds [16]. Dispersions of anisotropic (hard) colloidal particles have a more intricate phase behaviour [18]. Directional excluded volume interactions between anisotropic colloidal particles also give rise to entropy-driven phase transitions [19], which have been studied for lyotropic systems such as rod-like [20, 21] and platelet-like particles, [22] as well as for non-axisymmetric colloids [23, 24]. The communal entropy effects of directional interactions suffices to explain, for instance, the emergence of liquid-crystalline phases in suspensions of anisotropic colloidal particles [25]. Depletant addition to a system of anisotropic particles induces effective depletion attraction patches [26] because the depletion attraction is stronger for larger overlap of depletion zones. The effects of entropic patchiness in lyotropic systems have received increasing attention [27].

The specific colloid–polymer affinity also modulates the effective colloid–colloid interaction [28]. Whenever there is some polymer adhesion at the colloidal surface but this attraction does not suffice to compensate the configurational penalty for polymer adsorption, there is a non-negligible polymer concentration near the colloidal surface. This happens for instance when PDMS polymers in cyclohexane are close to silica surfaces [29]. However, classical depletion models assume a negligible polymer concentration at the colloidal surface. Attention has been paid to understand the effect of this *weak depletion* as compared to the classical depletion case [30, 31], also at high polymer concentrations [32]. When the colloid–polymer affinity is high enough, the entropic penalty for the polymer is balanced by an *enthalpic* gain near the colloidal surface, leading to (weak) polymer adsorption. This induces flocculation at low polymer concentration and steric stabilisation at higher concentrations [33]. In fact, a transition from polymer depletion to adsorption occurs with increasing colloid–polymer affinity.

A particular class of colloidal particle dispersions are association colloids [34]. In such systems the colloidal particles of interest are constituted by amphipathic building blocks, unimers, which in a selective solvent self-assemble into structures with colloidal size. Spherical micelles composed of block copolymers are an example of such association colloids [35]. Contrary to inorganic (hard) colloids, colloid-forming unimers can be in (thermodynamic) equilibrium with a bulk unimer concentration [36]. As the final equilibrium micelle properties are set via the unimers, micelle–micelle interactions can be modulated via the unimer properties. While a lot of attention has been paid to equilibrium micelle properties in terms of unimer composition and solvent conditions [37], less is known about the colloidal stability of micelles, and how addition of guest compounds, such as polymers, modulates micelle–micelle interactions.

1.2 Aim and Structure of This Thesis

This thesis constitutes a fundamental study on polymer-mediated interactions between spherical hard colloids, anisotropic hard colloids, and association spherical colloids. In order to understand the intricacies of colloid–polymer mixtures (CPMs), simplified versions of the real-life counterpart shall be studied. This enables extracting some of the key parameters governing the final properties of a colloidal suspension. A useful model is constructed over clear assumptions, and shall recover real-life observations; in return such a model may be used to delineate new experimental areas of interest. A textbook example of a complex colloid–polymer mixture is paint [38]. The particular shape of the pigment, as well as the specific interaction between polymers (binders) and pigments, play a role in the properties of the final product. Hence, fundamental understanding of how these two parameters (pigment shape and pigment–binder interaction) influence the phase stability of a paint suspension is not only of fundamental, but also of practical interest. Through this thesis, selected CPMs are studied putting emphasis to different key aspects governing their stability. The insights gained are compared, when possible, with experimental observations.

We employ available theoretical and computational tools as a starting point. If possible, the stable phases present for the CPM of interest at a given set of system parameters are cast into phase diagrams. Alternatively (or complementarily), an assessment of the colloidal stability is inferred via the second (osmotic) virial coefficient B_2. The effects of direct colloid–colloid interactions beyond their hard core on the stability of CPMs are studied in Chap. 2, while in Chap. 3 we re-examine and improve the solid description of a well-established theory for CPMs. In Chap. 4 the effect of varying colloid–polymer affinity on the stability of CPMs is addressed. These three chapters constitute the first block of this work: spherical colloids containing a hard core. The second block deals with anisotropic hard colloids: the rather rich phase behaviour of superball–depletant (Chap. 5) and platelet–depletant (Chap. 6) suspensions is studied. Finally, in Chaps. 7 and 8 we investigate the interactions between spherical diblock copolymer micelles and how addition of a second homopolymer species modulates them. These two chapters constitute the third block: spherical association colloids. In Chaps. 2, 3, 5, and 6 the depletion agent is considered 'ideal', whilst in Chaps. 4 and 8 the polymeric nature of the homopolymer added to the colloidal suspension is taken into account.

1.3 Common Methodology

To avoid repetition, this section serves as a common background for some concepts and tools used in this thesis. Firstly, we introduce the pair potentials used and the calculation of the second (osmotic) virial coefficient. The three main tools employed are also summarised here: Tang's first order mean spherical approximation (FMSA) for the hard-core Yukawa potential, Lekkerkerker's free volume theory (FVT), and Scheutjens–Fleer self-consistent mean-field computations (SCF) for micelle formation and micelle–micelle interactions. Specific, chapter-dependent methods are introduced when needed.

1.3.1 HCY, AOV, and PS Potentials

Within this thesis, the hard-core Yukawa (HCY) pair interaction [39] is frequently used either to model direct colloid–colloid interactions or as a fitting model for the interactions obtained via other approaches. The HCY potential mimics a wide range of interactions between *spherical* particles, since both the range and strength can be tuned. It could represent, for instance, a screened double layer repulsion or a Van der Waals attraction between the colloids [40]. It is convenient to work with the dimensionless distance between the centres of two colloidal spheres $\tilde{r} \equiv r/\sigma$, with r the centre–to–centre distance and σ the colloidal diameter ($\sigma = 2R$, with R the colloidal sphere radius). A tilde over the quantities is used to indicate dimensionless units. The relative range of the Yukawa interaction is characterised by $q_{\mathrm{Y}} = 2/(\hat{\kappa}\sigma)$, with $\hat{\kappa}$ the screening parameter (screening length $\hat{\kappa}^{-1}$). The HCY pair potential between colloidal spheres is written in terms of $\tilde{r}$, q_{Y} and ϵ, and normalised by $\beta = 1/(k_B T)$ (with k_B the Boltzmann's constant and T the absolute temperature):

$$\beta W_{\mathrm{HCY}} = \begin{cases} \infty & , \tilde{r} < 1 \\ -\frac{\beta\epsilon}{\tilde{r}} \exp\left[-\frac{2}{q_{\mathrm{Y}}}(\tilde{r} - 1)\right] & , \tilde{r} \geq 1. \end{cases} \tag{1.1}$$

The strength of the Yukawa potential is set via ϵ, defined such that $\epsilon > 0$ implies a HCY attraction and $\epsilon < 0$ a HCY repulsion. For $\epsilon = 0$ the HCY reduces to the HS interaction.

Following the classical works by Asakura, Oosawa [12, 13], and Vrij [14], the depletion-induced pair interaction between HSs can be described via an effective pair potential. We consider depletants with a radius $|\delta|$. Within this thesis, we consider the simplest depletant model in the AOV pair interaction. Depletants are described as penetrable hard spheres (PHSs): they do not interact with each other, but have a hard core repulsive interaction with the colloidal spheres. Thus, for PHSs as depletants, the thickness of the depletion zone is $|\delta| \equiv R_{\mathrm{g}}$, and $|\delta|$ is the PHS radius [15]. Note that in general δ refers to the adsorption thickness, and may be used to assess whether a polymer is depleted ($\delta < 0$) or adsorbed ($\delta > 0$) from/onto the colloidal particle

(Chaps. 4 and 8). For the ease of notation, in chapters where PHSs are considered we simply term the PHS radius as δ (Chaps. 2, 3, 5 and 6). The PHS approximation can be used to describe depletion effects in dilute polymer solutions: they mimic ideal polymer chains [15]. The depletion pair potential between HSs due to PHSs is the product of the overlap of depletion zones at a distance r times the bulk osmotic pressure of the ideal depletants [14]:

$$\beta W_{\mathrm{AOV}} = \begin{cases} \infty & , \tilde{r} < 1 \\ -\phi_{\mathrm{d}}^{\mathrm{R}} \left(\frac{1}{q} + 1\right)^3 \left[1 - \frac{3}{2}\frac{\tilde{r}}{q+1} + \frac{1}{2}\left(\frac{\tilde{r}}{q+1}\right)^3\right] & , 1 \leq \tilde{r} \leq 1+q \\ 0 & , \tilde{r} > 1+q, \end{cases} \quad (1.2)$$

where

$$q = \frac{2|\delta|}{\sigma} \quad (1.3)$$

is the relative size of the depletant, and where $\phi_{\mathrm{d}}^{\mathrm{R}}$ is the volume fraction of PHSs in the bulk. When $\phi_{\mathrm{d}}^{\mathrm{R}} = 0$ the AOV potential reduces, also, to the HS one. This pair potential does not account for the multi-body nature of the depletion interaction [41]. We finally note that the contact potential $[W(r = \sigma)]$ of the AOV interaction depends both on $\phi_{\mathrm{d}}^{\mathrm{R}}$ and q:

$$W_{\mathrm{AOV}}(r = \sigma) = -\frac{3 + 2q}{2q}\phi_{\mathrm{d}}^{\mathrm{R}}. \quad (1.4)$$

In Fig. 1.1, illustrative pair potentials are presented. On top of the AOV and HCY pair interactions, a penetrable sphere (PS) potential [42] is also presented:

$$\beta W_{\mathrm{PS}} = \begin{cases} \zeta & , \tilde{r} < 1 \\ 0 & , \tilde{r} > 1. \end{cases} \quad (1.5)$$

The PHS–PHS interaction is a particular case of the PS potential, for which $\zeta = 0$. If $\zeta = \infty$, the HS–HS interaction is recovered. The range of the interactions considered can be recognised by the separation distance at which the AOV interaction vanishes. For the HCY interaction, q_{Y} marks the value at which the interaction has decayed to $e^2(q_{\mathrm{Y}} + 1)$ its contact value ϵ.

1.3.2 Second Virial Coefficient

From the pair potentials, the second virial coefficient (B_2) can be extracted to assess the stability of the colloidal suspension [43, 44]. For any form of the pair interaction

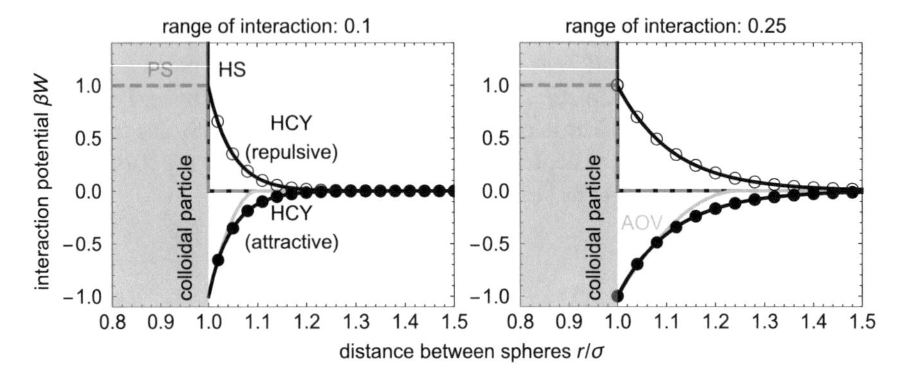

Fig. 1.1 Illustrative pair potentials for the AOV (grey) and HCY (circles) potentials considered in this thesis. For simplicity, all contact potentials are considered such that $|W(r=\sigma)| = 1$. An example PS potential (grey dashed) is also shown

$W(r)$, the second virial coefficient follows as:

$$\frac{B_2}{v_c} = 12 \int\limits_{\tilde{r}=0}^{\tilde{r}=\infty} \tilde{r}^2(1 - \exp[-\beta W(\tilde{r})])\mathrm{d}\tilde{r}, \tag{1.6}$$

where v_c is the colloidal particle volume. If the pair potential contains a hard core contribution, the equation above reads:

$$\frac{B_2}{v_c} = 4 + 12 \int\limits_{\tilde{r}=1}^{\tilde{r}=\infty} \tilde{r}^2(1 - \exp[-\beta W(\tilde{r})])\mathrm{d}\tilde{r}. \tag{1.7}$$

From Eq. (1.7) it follows that for HSs $B_2 = 4v_c$ [16]. Colloidal suspensions with B_2 above the HS limit are expected to be stable. If colloidal particles attract each other, $B_2 < 4v_c$. Whenever $B_2 \le -6v_c$ gas–liquid phase separation of the colloidal suspension is expected [45]. The latter is known as the Vliegenthart–Lekkerkerker (VL) criterion. The VL criterion can be regarded as a particular case of Noro and Frenkel's extended law of corresponding states [46]. In short, this law states that B_2 is useful to estimate the onset of demixing. For a colloidal suspension of sticky hard spheres (short-ranged attraction), the onset of phase separation is $B_2 \approx -4.9v_c$ [47]; slightly higher than, but close to the value, specified by the VL criterion. In Part III we use both the HS and the VL B_2-values as an *indicative* of the colloidal stability of a micellar suspension. The B_2-value dependence on system parameters may be used for sketching a *state diagram*.

In Fig. 1.2, illustrative B_2-values are presented for the potentials considered. We briefly address how strong a potential may be in order to recover $B_2 = 4v_c$ due to contributions *within* the colloidal domain. Consider the PS potential defined in the previous subsection. The B_2 of this potential ($B_2^{\mathrm{PS}}/v_c = 4\left[1 - e^{-\beta\zeta}\right]$) converges

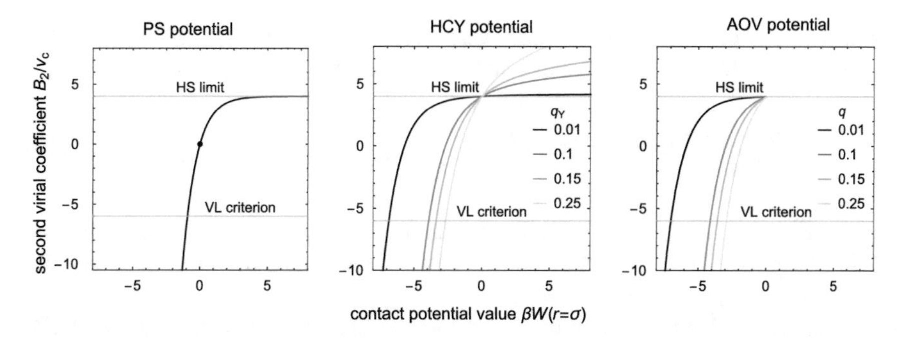

Fig. 1.2 Normalised second virial coefficient as a function of the contact value of the potential considered $[W(r = \sigma)]$ for the indicated pair potentials

towards $4v_c$ already for $\beta\epsilon \approx 5$ (see leftmost panel of Fig. 1.2). Hence, for any pair interaction with $W(r = \sigma) \approx 5k_BT$ or larger and *increasing* in $r \leq \sigma$, Eq. (1.7) remains (approximately) valid. This simple mental experiment shows that B_2 must be interpreted carefully: even if B_2 is close to the HS value, it does not imply that the particles behave as HSs. In fact, the fluid–solid coexistence region for PSs with $\beta\epsilon = 5$ occurs at volume fractions $\phi_c \in \{0.8, 1.0\}$ [42]: far above the coexistence region for HSs, $\phi_c \in \{0.49, 0.55\}$ [48].

For purely repulsive HCY potentials, B_2 increases with respect to the HS-value and is below the HS case for attractive HCY cases. For the AOV pair-interaction, $W_{AOV}(r = \sigma) < 0$ whenever the depletant concentration ϕ_d^R is finite. Consequently, $B_2 \leq 4v_c$ for all $W(r = \sigma)$.

1.3.3 FMSA: Closed Expressions for Interacting Colloids

Conveniently, from the HCY potential one can extract approximate (yet accurate) analytical thermodynamic expressions for the free energy of a HCY-interacting colloidal suspension (within certain limits) [49]. The free energy of a dispersion of colloidal spheres (F_c) interacting via hard-core Yukawa (F_{HCY}) is described as consisting of a hard core plus an additional Yukawa contribution [39, 50]:

$$\frac{F_c v_c}{k_B T V} \equiv \widetilde{F}_k^{\mathrm{HCY}} = \widetilde{F}_k^{\mathrm{HS}} + \widetilde{F}_{\mathrm{Y}}, \tag{1.8}$$

where V is the volume of the system considered, and k denotes the phase-state (fluid or solid). The pure HS contributions to the free energy ($\widetilde{F}_k^{\mathrm{HS}}$) are well-known [51, 52]. For a fluid of HSs, an accurate expression up to colloid volume fractions $\phi_c \approx 0.5$ follows from the Carnahan-Starling (CS) [52] equation of state (EOS):

$$\widetilde{F}_{\mathrm{fluid}}^{\mathrm{HS}} = \phi_c \left(\ln \frac{\phi_c \Lambda_{\mathrm{B}}^3}{v_c} - 1 \right) + \frac{4\phi_c^2 - 3\phi_c^3}{(1 - \phi_c)^2}, \tag{1.9}$$

where Λ_B is the de Broglie thermal wavelength. For a face-centred cubic (FCC) crystalline solid phase the Lennard–Jones–Devonshire (LJD) [51] EOS reads:

$$\widetilde{F}_{\text{solid}}^{\text{HS}} = 2.1306\phi_c + 3\phi_c \ln\left(\frac{\phi_c}{1-\phi_c/\phi_c^{\text{cp}}}\right) + \phi_c \ln(\Lambda_B^3/v_c), \tag{1.10}$$

where $\phi_c^{\text{cp}} = \pi/(3\sqrt{2}) \approx 0.74$ is the volume fraction of HSs at close packing. The value 2.1306 has been collected from Monte Carlo simulations of the pure HS system, [53] but is fairly close to the LJD solution.

Tang et al. [49] derived an expression for the free energy of a collection HCY spheres via a first-order mean spherical approximation (FMSA). The HCY potential allows an analytical solution of Ornstein–Zernike integral upon using the mean spherical closure approximation in Laplace space. This leads to analytical expressions for the radial distribution function and the direct correlation function up to first order in inverse temperature. The results provide a closed expression for F_Y in Eq. (1.8). Tang's approach can be extended to Multi-Yukawa potentials, and has been successfully applied to study the interactions between charged colloidal particles [54] and to predict multi-body properties of particles interacting through a Lennard–Jones pair interaction [39]. This Yukawa contribution to the free energy can be written in a Van der Waals form [39, 50]:

$$\widetilde{F}_Y = -\gamma_Y \phi_c^2,$$

where the Van der Waals parameter γ_Y is not a constant but reads:

$$\gamma_Y = \gamma_1 \beta\epsilon + \gamma_2 (\beta\epsilon)^2, \tag{1.11}$$

in which the functions γ_1 and γ_2 can be expressed in terms of the auxiliary functions L_Y and Q_Y:

$$\gamma_1 = \frac{3q_Y^2 L_Y}{(1-\phi_c)^2(1+Q_Y)} \quad , \quad \gamma_2 = \frac{3q_Y}{2(1+Q_Y)^4},$$

where

$$L_Y = 1 + 2/q_Y + \phi_c(2 + 1/q_Y),$$

and

$$Q_Y = \phi_c \frac{6(1-\phi_c)q_Y + 9\phi_c q_Y^2 - 3q_Y^3[1 + 2\phi_c - L_Y \exp(-2/q_Y)]}{2(1-\phi_c)^2}. \tag{1.12}$$

The osmotic pressure Π and chemical potential μ of HCY-interacting spheres follow from standard thermodynamic relations:

$$\beta\mu \equiv \widetilde{\mu} = \left(\frac{\partial \widetilde{F}_{\mathrm{c}}}{\partial \phi_{\mathrm{c}}}\right)_{T,V} \quad ; \quad \beta\Pi v_{\mathrm{c}} \equiv \widetilde{\Pi} = \phi_{\mathrm{c}}\widetilde{\mu} - \widetilde{F}_{\mathrm{c}}, \tag{1.13}$$

The relations in Eq. (1.13) apply in general, not only for interacting HCY spheres. Calculation of coexistence binodals and characteristic phase points from the FMSA EOS follow the same reasoning explained in the next Section. Further improvements of Tang's FMSA have been proposed [55, 56], yet lacking the simple and tractable closed forms presented here.

1.3.4 FVT for PHS-Depletants

We account for mixtures of hard colloidal particles plus non-adsorbing polymers in a semi-grand canonical fashion via the free volume theory (FVT) developed by Lekkerkerker et al. in the 1990s [15, 57, 58]. Within FVT, the colloid–polymer mixture (CPM, the system S of interest) is considered to be in equilibrium with a reservoir (R) of polymers. In R and S the solvent is treated as background. System and reservoir are connected through a membrane permeable for the polymers and the common solvent, but impermeable for the colloidal particles. A sketch of this osmotic equilibrium approach is given in Fig. 1.3 for a cuboid–polymer mixture. The key quantity relating the polymer concentrations in R and S is the free volume fraction available for depletants in the system α. Following the original ideas of FVT, we assume that α is independent of the chemical potential of the depletants in R. Furthermore, we take the simplest model for polymeric depletants, namely the PHS model introduced in Sect. 1.3.1. The FVT approach results in the following (normalised) expression for the semi-grand potential of the system:

$$\widetilde{\Omega} = \frac{\beta\Omega v_{\mathrm{c}}}{V} = \widetilde{F}_{\mathrm{c}} - \widetilde{\Pi}_{\mathrm{d}}^{\mathrm{R}}\alpha\frac{v_{\mathrm{c}}}{v_{\mathrm{d}}}, \tag{1.14}$$

with V the volume of the system, $\widetilde{\Pi}_{\mathrm{d}}^{\mathrm{R}} = \beta\Pi_{\mathrm{d}}^{\mathrm{R}}v_{\mathrm{d}}$ the reduced depletant osmotic pressure in R, and v_{d} the volume of the depletant ($v_{\mathrm{d}} = 4\pi|\delta|^3/3$). Since the depletants are considered to behave ideally, the (reduced) osmotic pressure in the R is simply given by Van 't Hoff's law:

$$\widetilde{\Pi}_{\mathrm{d}}^{\mathrm{R}} = \beta\Pi_{\mathrm{d}}^{\mathrm{R}}v_{\mathrm{d}} = \phi_{\mathrm{d}}^{\mathrm{R}}.$$

The depletant concentration in the system follows from:

$$\phi_{\mathrm{d}}^{\mathrm{S}} = \alpha\phi_{\mathrm{d}}^{\mathrm{R}}.$$

From the semi-grand potential, the chemical potential and osmotic pressure of the CPM of interest follow similarly as in Eq. (1.13) from standard thermodynamic relations:

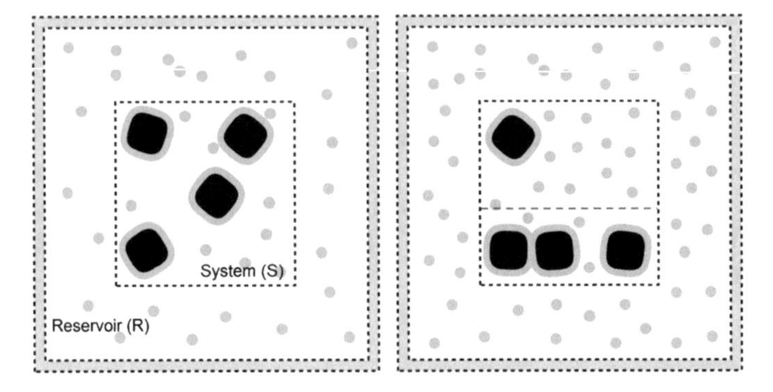

Fig. 1.3 Schematic representation of the free volume theory construction for colloid–depletant mixtures. Colloidal cuboids are indicated as big black particles; the corresponding depletion zones surrounding them are indicated in grey. The system of interest (centre) is surrounded by a reservoir containing only depletants (grey particles). Left panel: there is only one stable phase. Right panel: at high enough depletant concentration the system of interest is phase-separated into a cuboid-rich and a dilute cuboid phase

$$\beta\mu \equiv \widetilde{\mu} = \left(\frac{\partial \widetilde{\Omega}}{\partial \phi_c}\right)_{T,V,N_d^R} \quad ; \quad \beta \Pi v_c \equiv \widetilde{\Pi} = \phi_c \widetilde{\mu} - \widetilde{\Omega}, \tag{1.15}$$

with N_d^R the number of depletants in R. Using these quantities, phase coexistences follow from:

$$\widetilde{\mu}_i = \widetilde{\mu}_j = \cdots \quad \text{and} \quad \widetilde{\Pi}_i = \widetilde{\Pi}_j = \cdots, \tag{1.16}$$

where i and j denote the two (or more) coexisting phases of kind i, j. Coexistence between three phases corresponds to a triple point (TP). Four-phase coexistence is denoted via a quadruple point (QP). Colloidal systems may exhibit isostructural phase coexistence (such as gas–liquid equilibrium) when attractive interactions between particles are present [50, 58]. In such a case, the low-density phase will be entropically favourable and the high-density phase is stabilized by attractive interactions between the particles. Commonly, the limit of isostructural phase coexistence is defined via the critical point (CP), usally calculated via:

$$\frac{\partial \widetilde{\mu}_i}{\partial \phi_c} = \frac{\partial^2 \widetilde{\mu}_i}{\partial^2 \phi_c} = 0 \quad \text{and} \quad \frac{\partial \widetilde{\Pi}_i}{\partial \phi_c} = \frac{\partial^2 \widetilde{\Pi}_i}{\partial^2 \phi_c} = 0. \tag{1.17}$$

Whenever a phase state has a CP, an isostructural phase coexistence can take place, which can be stable or metastable. The transition from a potentially stable to a metastable isostructural phase coexistence is defined by the critical end point (CEP), where the CP and the TP (or QP) of the corresponding isostructural coexistences merge [50, 59]. These conditions enable to determine the topology of the phase diagrams as a function of the system parameters of the CPM of interest.

The only remaining unknown parameter in Eq. (1.14) is the free volume fraction for depletants in the system α. Widom's insertion theorem [60] relates the free volume fraction α to the work ω required to bring a depletant from R to S via:

$$\alpha = \frac{\langle V_{\mathrm{free}} \rangle_{\mathrm{o}}}{V} = e^{-\beta\omega}, \tag{1.18}$$

where $\langle V_{\mathrm{free}} \rangle_{\mathrm{o}}$ is the average free volume for depletants in the undistorted (depletant-free) system. This work ω is approximated using Scaled Particle Theory (SPT) [61, 62], by connecting the limits of inserting a very small depletant and a very big depletant in the system S of interest, followed by scaling back to the actual size of the depletant. In this thesis, we only consider spherical depletants. Hence, a single scaling factor (λ) enables to express this work by combining the limiting results for $\lambda \to 0$ and $\lambda \to \infty$:

$$\omega(\lambda) = \underbrace{\omega(0) + \left.\frac{\partial \omega}{\partial \lambda}\right|_{\lambda=0} \lambda + \frac{1}{2} \left.\frac{\partial^2 \omega}{\partial \lambda^2}\right|_{\lambda=0} \lambda^2}_{\lambda \ll 1} + \underbrace{v_{\mathrm{d}} \Pi_k^{\mathrm{o}}}_{\lambda \gg 1}, \tag{1.19}$$

where Π_k^{o} is the osmotic pressure of the depletant-free system (in a phase k). In the small depletant insertion limit ($\lambda \ll 1$) there is no overlap of depletion zones: $\alpha \to [1 - \phi_{\mathrm{c}} v_{\mathrm{exc}}(\lambda)/v_{\mathrm{c}}]$, with v_{exc} the excluded volume between the hard colloid of interest and a sphere. Equation (1.18) then allows writing $\omega(\lambda \ll 1)$ as:

$$\beta\omega(\lambda \ll 1) = -\ln\left[1 - \phi_{\mathrm{c}} \left(\frac{v_{\mathrm{exc}}(\lambda)}{v_{\mathrm{c}}}\right)\right], \tag{1.20}$$

where the scaled depletion volume is obtained by scaling the depletant size: $|\delta| \to \lambda|\delta|$. For big depletants ($\lambda \to \infty$) we assume that the insertion work ω is the work required to create a cavity with the size of the depletant in the system. We use normalised units also in the big-depletant limit for convenience, thus:

$$\beta\omega(\lambda \gg 1) = \frac{v_{\mathrm{d}}(\lambda)}{v_{\mathrm{c}}} \widetilde{\Pi}_k^{\mathrm{o}}. \tag{1.21}$$

By combining Eqs. (1.19) and (1.21), and inserting $\lambda = 1$ a general expression for ω is derived:

$$\begin{aligned} \beta\omega = & \underbrace{-\ln(1 - \phi_{\mathrm{c}})}_{\text{point depletant}} \\ & + \underbrace{y \left.\frac{\partial \widetilde{v}_{\mathrm{exc}}(\lambda)}{\partial \lambda}\right|_{\lambda=0} + \frac{y}{2}\left(\left.\frac{\partial \widetilde{v}_{\mathrm{exc}}(\lambda)}{\partial \lambda}\right|_{\lambda=0}\right)^2 + \frac{y}{2} \left.\frac{\partial^2 \widetilde{v}_{\mathrm{exc}}(\lambda)}{\partial \lambda^2}\right|_{\lambda=0}}_{\text{colloidal shape-dependent term, } Q_{\mathrm{s}}} \\ & + \underbrace{\frac{v_{\mathrm{d}}}{v_{\mathrm{c}}} \widetilde{\Pi}_k^{\mathrm{o}}}_{\text{cavity limit}}, \end{aligned} \tag{1.22}$$

with $\widetilde{v}_{\mathrm{exc}} = v_{\mathrm{exc}}/v_{\mathrm{c}}$ and

$$y = \frac{\phi_{\mathrm{c}}}{1-\phi_{\mathrm{c}}} \quad .$$

Inserting Eq. (1.22) into Eq. (1.14) using the relation given in Eq. (1.18) yields the general expression:

$$\widetilde{\Omega} = \widetilde{F}_{\mathrm{c}} - \frac{v_{\mathrm{c}}}{v_{\mathrm{d}}} \widetilde{\Pi}_{\mathrm{d}}^{\mathrm{R}} (1-\phi_{\mathrm{c}}) \exp[-Q_{\mathrm{s}}] \exp\left[-\frac{v_{\mathrm{d}}}{v_{\mathrm{c}}} \widetilde{\Pi}_{\mathrm{c}}^{\mathrm{o}}\right]. \tag{1.23}$$

Hence, provided the particular expression for v_{exc} and the EOS of the depletant-free system are known, determination of phase coexistence is straightforward. All results presented for the FVT approaches can be easily extended towards polymers in Θ and good solvent conditions at elevated concentrations [15]. Independent studies have shown that, for instance, the phase behaviour of anisotropic colloid–polymer mixtures are not dramatically sensitive to the specific nature of the depletant [63]. Therefore it is expected this approach can provide insight into a wide range of mixtures of colloids plus depletants while being tractable.

1.3.5 SCF for Micelle–Micelle Interactions

In the last two chapters of this thesis we focus on spherical micelles, particularly on micelle–micelle interactions. Specifically, we studied diblock copolymer micelle solutions. We used the Scheutjens–Fleer self-consistent lattice theory (SCF) [64–66] to numerically resolve micelle–micelle interactions, employing the `sfbox` software developed by prof. F.A.M. Leermakers at Wageningen University. This software was also used in the computations in Chap. 4. It is based upon Flory–Huggins mean-field theory, [67] but with concentration gradients following discrete versions of the Edwards equation accounting for mean-field polymer propagations [68]. The interdependence of the (segment) potentials u_k and the volume fractions ϕ_k for each component k in the system,

$$\phi_k(u_k) \leftrightarrow u_k(\phi_k) \quad ,$$

is the core idea of the self-consistent method [69]. Provided an user-defined starting configuration for the components in the lattice, the free energy (F) of the lattice is optimised in a self-consistent fashion.

Explain more about SCF here?

The SCF routine provides concentration profiles from which other quantities may be derived. For instance, the method used to calculate the hydrodynamic size of self-assembled structures from SCF density profiles follows the work of Scheutjens et al. [66, 70, 71] It is based upon applying the Debye-Brinkman equation to the SCF-computed polymer segment density profile. The core idea is to relate the (theoretical) solvent velocity profile $[v(z)]$ to the polymer segment concentration profile $\phi_\mathrm{p}(z)$. In each layer at position z, the quantity $\bar{\alpha}(z) = v(z)/v'(z)$ can be related to the segment concentration profile via:

$$\bar{\alpha} = \frac{\bar{\rho}(z)\tanh[\bar{\rho}(z)]^{-1} + \bar{\alpha}(z-1)}{1 + \bar{\alpha}(z-1)\bar{\rho}^{-1}(z)\tanh[\bar{\rho}(z)]^{-1}}, \tag{1.24}$$

where $\bar{\rho}(z)$ is a normalised polymer segment concentration profile with respect to the bulk concentration:

$$\bar{\rho} = \sqrt{\frac{1 - \phi_\mathrm{p}(z) + \phi_\mathrm{p}^\mathrm{bulk}}{\phi_\mathrm{p}(z) - \phi_\mathrm{p}^\mathrm{bulk}}}. \tag{1.25}$$

The hydrodynamic radius (in lattice units) hence relates to the outermost layer at which there is no more change in the relative solvent velocity.

Within SCF, the boundary conditions and the lattice geometry need to be specified. For spherical micelles we consider two different lattice types, namely a spherical lattice with concentration gradients in one direction and a cylindrical lattice with concentration gradients in two directions. Mirror boundary conditions are set for all boundaries [72]. A spherical lattice is defined as shells from the centre ($z = 0$) of the lattice up to $z = N_\mathrm{lat}$ (N_lat beign the number of lattice sites). The first lattice layer corresponds to the centre of the spherical micelle. As we focus on conditions where spherical micelles are preferred over other self-assembled morphologies, a spherical lattice is used in most of our calculations. A cylindrical lattice is defined by a grid of N_lat^r sites in the radial coordinate, and N_lat^y sites in the longitudinal coordinate, and it is used in Chap. 7 simply to assess the accuracy of the calculation of the micelle–micelle interactions using the spherical lattice.

Due to the mirror conditions imposed, a micelle is formed in the presence of K surrounding ones. The distance between the centres of two nearest neighbour micelles r defines the characteristic length scale involved in the calculation of micelle–micelle interactions. For the spherical lattice, $K = 12$ and $r = 2N_\mathrm{lat}$. For the cylindrical one, six cylindrical lattices are present around the simulated one in the radial direction ($K_r = 6$) while two span from the top and bottom of the radial mirror conditions (one from the upper and one from the lower boundaries of the lattice ($K_y = 2$). In this case, the micelle–micelle interaction calculation depends on how $N_\mathrm{lat}^{(k)}$ is varied ($k = \{r, y\}$). In the cylindrical lattice, micelles are formed in the centre of the radial axis of symmetry. This implies that the nearest neighbours are at distances $r_r = 2N_\mathrm{lat}^r$ (radial direction) and $r_y = N_\mathrm{lat}^y$ (longitudinal direction). Both for the spherical and cylindrical lattices a lattice constant $k = 3$ was used, following previous research on SCF applied to self-assembly [73].

The SCF approach is combined with small system thermodynamics to study the conditions under which the diblock copolymers form self-assembled morphologies [74, 75]. To find the equilibrium configuration, we compute the grand potential Ω of the system for a specific diblock copolymer as a function of the aggregation number g_p:

$$\Omega = F - \sum_{i=1}^{N_t} N_i \mu_i, \tag{1.26}$$

where μ_i and N_i are the chemical potential and (number) concentrations of the species i, and N_t is the number of components considered. This grand potential relates to the inhomogeneities in the system: in a pure solvent $\Omega = 0$. As diblock copolymers are added to the solution, Ω increases due to the contacts present between solution and solvophobic blocks. The appearance of the first thermodynamically stable micelle is marked by a maximum in Ω. If a micelle can form, Ω decreases with g_p, and at a given diblock concentration the condition $\Omega_{g_p \neq 0} = 0$ is met (with $\partial\Omega/\partial g_p < 0$) [34]. At this condition the block copolymers in the micelle are in equilibrium with free block copolymers in the bulk. Thus, the chemical potential of one copolymer in the micelle is equal to that in the bulk:

$$\mu_p^{\text{bulk}} = \mu_p^{\text{micelle}} \equiv \mu_p. \tag{1.27}$$

The work required to dissociate all polymers from the fully-grown self-assembled structure (equivalently, the energy gain of the diblocks upon micellisation) at a certain intermicelle distance r follows as [76]:

$$\omega(r) = g_p(r)\mu_p(r) + N_s(r)\mu_s(r) \quad . \tag{1.28}$$

Note that this work is the free energy of micellisation when $\Omega = 0$ [Eq. (1.26)]. A similar approach has been followed to study the phase behaviour of flower-like micelles [77]. If a homopolymer (G) is also added to the considered colloidal suspension (Chap. 8), we compute the work required to dissociate a micelle in its presence (the free energy difference when a second component is present in the system):

$$\omega(r) = g_p(r)\mu_p(r) + N_s(r)\mu_s(r) + g_G(r)\mu_G(r), \tag{1.29}$$

with g_G the excess number of guest homopolymer in the system. The SCF-provided equilibrium quantities enable us to estimate the micelle–micelle interaction as:

$$W(x) = \frac{2}{K}\left[\omega(r) - \omega(r = \infty)\right] \quad . \tag{1.30}$$

These effective interactions consider micelles composed of g_p polymers surrounded by K other micelles at a distance r, and are directly estimated from the SCF output. Our approach does not imply any *ad hoc* interaction between the micellar domains. Rather, they originate naturally from the equilibrium properties of micelles formed at a given r. From typical diffusion coefficient and micelle size values for block copolymer micelles [78, 79], the time required for a micelle to travel over its own size (the configurational relaxation time [1]) is of the order of μs. This time estimate applies independently of the relative block lengths. Remarkably, the unimer–micelle exchange *may* also be up to the μs time scale if the solvophilic block is significantly larger than the solvophobic one. If, on the contrary, the solvophobic and solvophilic blocks are of similar chain length, the unimer–micelle exchange rate drops dramatically [80, 81]. Such self-assembled morphologies are often termed 'frozen micelles'. These effects are linked to the energy barrier of removing a unimer from the micelle, which is smaller the larger the solvophilic block. Contrary to hard, non-associative particles, (spherical) association colloids *may* adapt their properties while interacting with each other.

Note

Numerical computations (e.g., phase diagram calculations, B_2 calculations), data processing (e.g., `sfbox` and Monte Carlo output data), and plots were conducted using *Wolfram Mathematica* [82]. *Mathematica* scripts are available from the author upon reasonable request.

References

1. A.P. Philipse, *Brownian Motion: Elements of Colloid Dynamics*, Undergraduate Lecture Notes in Physics (Springer International Publishing, New York, 2018). www.springer.com/gp/book/9783319980522
2. W.C.K. Poon, Science **304**, 830 (2004). https://doi.org/10.1126/science.1097964
3. A. Einstein, Ann. Phys. **324**, 371 (1905). https://doi.org/10.1002/andp.19063240208
4. J. Perrin, Ann. Chim. Phys. **18**, 5 (1909), http://www.citeulike.org/group/744/article/1059645
5. M.D. Haw, J. Phys.: Condens. Matter **14**, 7769 (2002). http://stacks.iop.org/0953-8984/14/i=33/a=315
6. E.J.W. Verwey, J.T.G. Overbeek, *Theory of the Stability of Lyophobic Colloids* (Elsevier, Amsterdam, 1948)
7. D.H. Everett, *Basic Principles of Colloid Science* (Royal Society of Chemistry, London, 1988)
8. J.N. Israelachvili, *Intermolecular and Surface Forces*, 3rd edn. (Academic Press, Amsterdam, 2011)
9. R.J. Hunter, *Foundations of Colloid Science* (Oxford University Press, Oxford, 2001)
10. A. Yethiraj, A. van Blaaderen, Nature **421**, 513 (2003). https://doi.org/10.1038/nature01328
11. J.A. Lewis, J. Am. Ceram. Soc. **83**, 2341 (2000). https://doi.org/10.1111/j.1151-2916.2000.tb01560.x
12. S. Asakura, F. Oosawa, J. Chem. Phys. **22**, 1255 (1954). https://doi.org/10.1063/1.1740347
13. S. Asakura, F. Oosawa, J. Polym. Sci. **33**, 183 (1958). https://doi.org/10.1002/pol.1958.1203312618
14. A. Vrij, Pure Appl. Chem. **48**, 471 (1976). https://doi.org/10.1351/pac197648040471
15. H.N.W. Lekkerkerker, R. Tuinier, *Colloids and the Depletion Interaction* (Springer, Heidelberg, 2011)
16. A. Mulero, *Theory and Simulations of Hard-Sphere Fluids and Related Sytem* (Springer, Heidelberg, 2008)
17. R.P.A. Dullens, Soft Matter **2**, 805 (2006), http://xlink.rsc.org/?DOI=b607017e
18. L. Onsager, Ann. N. Y. Acad. Sci. **51**, 627 (1949). https://doi.org/10.1111/j.1749-6632.1949.tb27296.x
19. D. Frenkel, Phys. A (Amsterdam, Neth.) **263**, 26 (1999), https://doi.org/10.1016/S0378-4371(98)00501-9
20. G.J. Vroege, H.N.W. Lekkerkerker, Rep. Prog. Phys. **55**, 1241 (1992), http://iopscience.iop.org/article/10.1088/0034-4885/55/8/003/pdf

21. S. Varga, A. Galindo, G. Jackson, Mol. Phys. **101**, 817 (2003). https://doi.org/10.1080/0026897021000037654
22. J.A.C. Veerman, D. Frenkel, Phys. Rev. A **45**, 5632 (1992), https://journals.aps.org/pra/abstract/10.1103/PhysRevA.45.5632
23. A. Haji-Akbari, M. Engel, A.S. Keys, X. Zheng, R.G. Petschek, P. Palffy-Muhoray, S.C. Glotzer, Nature **462**, 773 (2009). https://www.nature.com/articles/nature08641
24. A.P. Gantapara, J. de Graaf, R. van Roij, M. Dijkstra, J. Chem. Phys. **142**, 054904 (2015). https://aip.scitation.org/doi/10.1063/1.4906753
25. M. Dijkstra, in *Advances in Chemical Physics*, vol. 156, ed. by S.A. Rice A.R. Dinner (Wiley, NewYork, 2014) Chap. 2. https://onlinelibrary.wiley.com/doi/10.1002/9781118949702.ch2
26. G. van Anders, N.K. Ahmed, R. Smith, M. Engel, S.C. Glotzer, ACS Nano **8**, 931 (2014). https://pubs.acs.org/doi/10.1021/nn4057353
27. A.S. Karas, J. Glaser, S.C. Glotzer, Soft Matter **12**, 5199 (2016). https://pubs.rsc.org/en/content/articlelanding/2016/sm/c6sm00620e#!divAbstract
28. X. Xing, L. Hua, T. Ngai, Curr. Opin. Colloid Interface Sci. **20**, 54 (2015). https://www.sciencedirect.com/science/article/pii/S1359029414001459
29. R. Tuinier, S. Ouhajji, P. Linse, Eur. Phys. J. E **39** (2016). https://link.springer.com/article/10.1140
30. W.K. Wijting, W. Knoben, N.A.M. Besseling, F.A.M. Leermakers, M.A. Cohen Stuart, Phys. Chem. Chem. Phys. **6**, 4432 (2004). https://pubs.rsc.org/en/Content/ArticleLanding/2004/CP/b404030a#!divAbstract
31. S. Ouhajji, T. Nylander, L. Piculell, R. Tuinier, P. Linse, A.P. Philipse, Soft Matter **12**, 3963 (2016). https://pubs.rsc.org/en/Content/ArticleLanding/2016/SM/C5SM02892B#!divAbstract
32. J.B. Hooper, K.S. Schweizer, Macromolecules **39**, 5133 (2006). https://pubs.acs.org/doi/10.1021/ma060577m
33. J. Gregory, S. Barany, Adv. Colloid Interface Sci. **169**, 1 (2011). https://www.sciencedirect.com/science/article/pii/S0001868611001229
34. F.A.M. Leermakers, J.C. Eriksson, J. Lyklema, in *Soft Colloids*, ed. by J. Lyklema. Fundamentals of Interface and Colloid Science, vol. 5 (Academic Press, Cambridge, 2005) Chap. 4. https://www.sciencedirect.com/bookseries/fundamentals-of-interface-and-colloid-science/vol/5/suppl/C
35. I. Hamley, *Block Copolymers in Solution: Fundamentals and Applications* (Wiley, New York, 2005)
36. C.B.E. Guerin, I. Szleifer, Langmuir **15**, 7901 (1999)
37. Y. Mai, A. Eisenberg, Chem. Soc. Rev. **41**, 5969 (2012). https://pubs.rsc.org/en/content/articlelanding/2012/cs/c2cs35115c#!divAbstract
38. T. Tadros, *Colloids in Paints* (Wiley, New York, 2011)
39. Y. Tang, Y.-Z. Lin, Y.-G. Li, J. Chem. Phys. **122**, 184505 (2005). https://aip.scitation.org/doi/10.1063/1.1895720
40. A. Fortini, M. Dijkstra, R. Tuinier, J. Phys.: Condens. Matter **17**, 7783 (2005). http://stacks.iop.org/0953-8984/17/i=50/a=002
41. M. Dijkstra, R. van Roij, R. Roth, A. Fortini, Phys. Rev. E **73**, 041404 (2006). https://journals.aps.org/pre/abstract/10.1103/PhysRevE.73.041404
42. C.N. Likos, M. Watzlawek, H. Löwen, Phys. Rev. E **58**, 3135 (1998). https://journals.aps.org/pre/abstract/10.1103/PhysRevE.58.3135
43. M.L. Kurnaz, J.V. Maher, Phys. Rev. E **55**, 572 (1997). https://journals.aps.org/pre/abstract/10.1103/PhysRevE.55.572
44. A. Quigley, D. Williams, Eur. J. Pharm. Biopharm. **96**, 282 (2015). https://www.sciencedirect.com/science/article/pii/S0939641115003288
45. G.A. Vliegenthart, H.N.W. Lekkerkerker, J. Chem. Phys. **112**, 5364 (2000). https://aip.scitation.org/doi/10.1063/1.481106
46. M.G. Noro, D. Frenkel, J. Chem. Phys. **113**, 2941 (2000). https://aip.scitation.org/doi/10.1063/1.1288684

47. R. Fantoni, A. Giacometti, A. Santos, J. Chem. Phys. **142**, 224905 (2015). https://aip.scitation.org/doi/full/10.1063/1.4922263
48. W.G. Hoover, F.H. Ree, J. Chem. Phys. **49**, 3609 (1968). https://aip.scitation.org/doi/10.1063/1.1670641
49. Y. Tang, B.C. Lu, J. Chem. Phys. **99**, 9828 (1993). https://aip.scitation.org/doi/10.1063/1.465465
50. R. Tuinier, G.J. Fleer, J. Phys. Chem. B **110**, 20540 (2006). https://pubs.acs.org/doi/abs/10.1021/jp063650j
51. J.E. Lennard-Jones, A.F. Devonshire, Proc. R. Soc A **163**, 53 (1937). https://www.jstor.org/stable/97067?seq=1#page_scan_tab_contents
52. N.F. Carnahan, K.E. Starling, J. Chem. Phys. **51**, 635 (1969). https://aip.scitation.org/doi/10.1063/1.1672048
53. D. Frenkel, A.J.C. Ladd, J. Chem. Phys. **81**, 3188 (1984). https://aip.scitation.org/doi/10.1063/1.448024
54. Y.-Z. Lin, Y.-G. Li, J.D. Li, J. Mol. Liq. **125**, 29 (2006). https://www.sciencedirect.com/science/article/pii/S0167732205001674
55. S. Hlushak, S. Trokhymchuk, I. Nezbeda, Condens. Matter Phys. **14**, 33004 (2011). http://www.icmp.lviv.ua/journal/zbirnyk.67/33004/abstract.html
56. S. Hlushak, A. Trokhymchuk, Condens. Matter Phys. **15**, 23003 (2012). http://www.icmp.lviv.ua/journal/zbirnyk.70/23003/abstract.html
57. H.N.W. Lekkerkerker, Colloids Surf. **51**, 419 (1990). https://www.sciencedirect.com/science/article/abs/pii/016666229080156X
58. H.N.W. Lekkerkerker, W.C.K. Poon, P.N. Pusey, A. Stroobants, P.B. Warren, Europhys. Lett. **20**, 559 (1992). https://doi.org/10.1209/0295-5075/20/6/015
59. C.F. Tejero, A. Daanoun, H.N.W. Lekkerkerker, M. Baus, Phys. Rev. Lett. **73**, 752 (1994). https://journals.aps.org/prl/abstract/10.1103/PhysRevLett.73.752
60. B. Widom, J. Chem. Phys. **39**, 2808 (1963). https://aip.scitation.org/doi/10.1063/1.1734110
61. E. Helfand, H. Reiss, H.L. Frisch, J.L. Lebowitz, J. Chem. Phys. **33**, 1379 (1960). https://doi.org/10.1063/1.1731417
62. J.L. Lebowitz, E. Helfand, E. Praestgaard, J. Chem. Phys. **43**, 774 (1965). https://doi.org/10.1063/1.1696842
63. V.F.D. Peters, M. Vis, Á. González García, R. Tuinier, in preparation (n.a.)
64. J.M.H.M. Scheutjens, G.J. Fleer, J. Chem. Phys. **83**, 1619 (1979). https://pubs.acs.org/doi/abs/10.1021/j100475a012
65. J.M.H.M. Scheutjens, G.J. Fleer, J. Chem. Phys. **84**, 178 (1980). https://pubs.acs.org/doi/abs/10.1021/j100439a011
66. G.J. Fleer, M.A. Cohen Stuart, J.M. H.M. Scheutjens, T. Cosgrove, B. Vincent, *Polymers at interfaces* (Springer, Netherlands, 1998) pp. XX, 496
67. P.J. Flory, *Principles of Polymer Chemistry*, The George Fisher Baker Non-Resident Lectureship in Chemistry at Cornell University (Cornell University Press, New York, 1953)
68. G.J. Fleer, Adv. Colloid Interface Sci. **159**, 99 (2010). https://doi.org/10.1016/j.cis.2010.04.004
69. F.A.M. Leermakers, J. Sprakel, N.A.M. Besseling, P.A. Barneveld, Phys. Chem. Chem. Phys. **9**, 167 (2006). https://doi.org/10.1039/B613074G
70. M.A. Cohen Stuart, F.H. W.H. Waajen, T. Cosgrove, B. Vincent, T.L. Crowley, Macromolecules **17**, 1825 (1984). https://pubs.acs.org/doi/abs/10.1021/ma00139a035
71. J.M.H.M. Scheutjens, G.J. Fleer, M.A. Cohen Stuart, Colloids Surf. **21**, 285 (1986). https://doi.org/10.1016/0166-6622(86)80098-1
72. E. Hilz, F.A.M. Leermakers, A.W.P. Vermeer, Phys. Chem. Chem. Phys. **14**, 4917 (2012). https://doi.org/10.1039/C2CP40318H
73. J. Lyklema, ed., "Appendix 1 - self-consistent field modelling," in *booktitle Soft Colloids*, Fundamentals of Interface and Colloid Science, Vol. 5 (Academic Press, 2005) pp. A1.1 – A1.12http://www.sciencedirect.com/science/article/pii/S1874567905800133

74. T.L. Hill, *Thermodynamics of Small Systems, Parts I & II*, vol. 3 (WILEY-VCH, Germany, 1965). https://doi.org/10.1002/ijch.196500008
75. Y. Lauw, F.A.M. Leermakers, M.A. Cohen Stuart, J. Phys. Chem. B **110**, 465 (2006). https://doi.org/10.1021/jp053795a
76. R. Feynman, R. Leighton, M. Sands, *The Feynman Lectures on Physics*, vol. I, Chap. 14 (1965).http://www.feynmanlectures.caltech.edu/I_14.html
77. J. Sprakel, N.A.M. Besseling, M.A. Cohen Stuart, F.A.M. Leermakers, Eur. Phys. J. E **25**, 163 (2008). http://link.springer.com/10.1140/epje/i2007-10277-1
78. A. Kelarakis, V. Havredaki, X.-F. Yuan, Y.-W. Yang, C. Booth, J. Mater. Chem. **13**, 2779 (2003). http://xlink.rsc.org/?DOI=B304254E
79. A. Ianiro, J. Patterson, Á. González García, M.M.J. van Rijt, M.M.R.M. Hendrix, N.A.J.M. Sommerdijk, I.K. Voets, A.C.C. Esteves, R. Tuinier, J. Polym. Sci., Part B: Polym. Phys. **56**, 330 (2018). https://onlinelibrary.wiley.com/doi/full/10.1002/polb.24545
80. R. Lund, L. Willner, J. Stellbrink, A. Radulescu, D. Richter, Macromolecules **37**, 9984 (2004). http://pubs.acs.org/doi/abs/10.1021/ma035633n
81. J.G.J.L. Lebouille, L.F.W. Vleugels, A.A. Dias, F.A.M. Leermakers, M.A. Cohen Stuart, R. Tuinier, Eur. Phys. J. E **36** (2013). http://link.springer.com/10.1140/epje/i2013-13107-y
82. R. Wolfram, Mathematica, Version 11.3, (2018), note Champaign, IL

Part I
Spherical Colloids

Chapter 2
Tuning the Phase Diagram of Colloid–Polymer Mixtures

2.1 Introduction

When considering hard spheres (HSs, diameter σ) depleted by penetrable hard spheres (PHSs, diameter 2δ), two different regimes for depletion interaction can be distinguished in terms of the relative depletant size $q \equiv 2\delta/\sigma$. For $q \lesssim 0.15$, a collection of spheres whose interaction is mediated by PHS-depletants is pair-wise additive [1]. Otherwise, multiple overlap of depletion zones needs to be accounted for. The consequences of assuming pair-wise addition in the latter case have been analysed and discussed in detail [2, 3]. Describing the mixture by an effective pair potential approach has been widely applied [4, 5]. Contrary to the usually complex or computationally-based approaches followed, in this chapter we describe a simple (yet relatively accurate) method for predicting phase diagrams of colloid–polymer mixtures (CPMs), in which the direct colloid–colloid interactions are not too strong.

Much attention has been paid to the phase behaviour of HSs plus depletants when multiple overlap of depletion zones occur ($q \gtrsim 0.15$) [6–8]. However, limited theories are available for colloidal particles interacting beyond their hard-cores and added free non-adsorbing polymers (acting as depletants). Available approaches focused on short-range repulsive colloids [9–13]. In real systems, more involved (soft) interactions are often present. For example, in paint and food dispersions, particles are often charged and/or partially (de)stabilized by adsorbing polymers [14].

We show in this chapter how additional direct interactions between the colloidal particles modify the phase diagram of a CPM. The direct interactions between the spherical colloidal particles are described via hard-core Yukawa potentials [15] [HCY, see Eq. (1.1)]. The non-adsorbing polymers are treated as PHSs [16] (see Chap. 1). In Sect. 2.3, phase diagrams of HCY spheres in a sea of PHSs are presented and compared with Monte Carlo (MC) simulations. It is shown how additional (relatively weak) direct colloid–colloid interactions modify the phase coexistence landscape within the framework of free volume theory (FVT) [17].

© Springer Nature Switzerland AG 2019

Á. González García, *Polymer-Mediated Phase Stability of Colloids*,
Springer Theses, https://doi.org/10.1007/978-3-030-33683-7_2

2.2 FVT for Hard-Core Yukawa Colloids

As introduced in Chap. 1, the required FVT ingredients for colloidal particles mixed with PHSs are the equation(s) of state of the pure colloidal suspension and the excluded volume between a colloidal particle and a PHS. For a dispersion of HCY-interacting colloidal spheres, we use the closed expressions provided by the first order mean spherical approximation [15]. From the general expression given in Chap. 1:

$$\widetilde{\Omega} = \widetilde{F}_k^{\mathrm{HCY}} - q^{-3}\widetilde{\Pi}_{\mathrm{d}}^{\mathrm{R}}\alpha, \tag{2.1}$$

the free volume fraction for depletants in the system α follows as:

$$\alpha = (1 - \phi_{\mathrm{c}}) \exp[-Q_{\mathrm{s}}] \exp\left[-q^3\widetilde{\Pi}_k^{\mathrm{o}}\right], \tag{2.2}$$

where $\widetilde{\Pi}_k^{\mathrm{o}} \equiv \widetilde{\Pi}_k^{\mathrm{HCY}}$ follows from Eq. (1.8) using standard thermodynamic relations [see Eq. (1.13)]. The term Q_{s} reads as for hard-spheres mixed with PHSs, as it is assumed that the direct colloid–colloid interactions do not affect the depletion zones:

$$Q_{\mathrm{s}} = 3qy + \frac{y}{2}q^2 y\,[6 + 9y]\,, \tag{2.3}$$

where

$$y = \frac{\phi_{\mathrm{c}}}{1 - \phi_{\mathrm{c}}}. \tag{2.4}$$

Logically, Eq. (2.1) reduces to the original FVT expression when considering a HCY potential with $\epsilon = 0$. In Fig. 2.1 we present the osmotic pressure of a HCY fluid ($\widetilde{\Pi}_{\mathrm{F}}^{\mathrm{o}}$) and the free volume fraction α (insets) for PHSs for the system parameters indicated. Added Yukawa attractions lower the osmotic pressure $\widetilde{\Pi}_{\mathrm{F}}^{\mathrm{o}}$ of the colloidal dispersions, while repulsions increase $\widetilde{\Pi}_{\mathrm{F}}^{\mathrm{o}}$ with respect to the pure HS interaction. This can be explained by the fact that an additional weak repulsion increases the effective excluded volume of the colloidal particle [9]. An inverse effect is expected for weak attractions. This has a significant influence on the stability of the gas–liquid (G–L) and fluid–solid (F–S) coexistence regions, inducing different coexistence with respect to the HS case at given system parameters ($q, q_{\mathrm{Y}}, \epsilon$).

In the insets of Fig. 2.1, α is compared to the HS-case in a colloidal fluid state. As the depletants are considered as PHSs, α does not to depend on the depletant concentration ($\phi_{\mathrm{d}}^{\mathrm{R}}$). Varying q has a greater effect on α than tuning the HCY interaction. This naturally follows from the fact that all HCY contributions to α contain a q^3 pre-factor, whereas depletant contributions are of the order of q and q^2 [see Eq. (2.2)]. In all cases, the difference in α between the HS and the HCY cases is small, but (logically) increases with q and with increasing strength $|\epsilon|$ of the Yukawa interactions. Note also that α increases with respect to the HS case for repulsive HCY

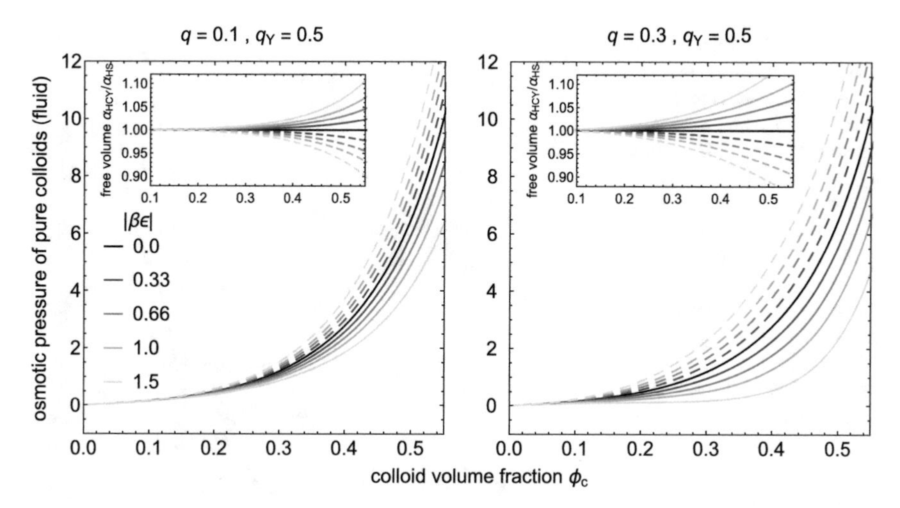

Fig. 2.1 Dimensionless osmotic pressure of the fluid phase of hard-core Yukawa (HCY) fluids for a collection of ϵ-values (black curve corresponds to $\epsilon = 0$; solid grey curves correspond to HCY attractions whilst dashed grey curves indicate HCY repulsions). Insets show the difference in free volume fraction with respect to the hard sphere case for the relative depletant–to–colloid size q indicated

interactions while it decreases for added HCY attractions [due to the negative sign of the osmotic pressure in Eq. (2.2)].

2.3 Results and Discussion

Firstly, we address the phase diagrams of HCY spheres mixed with PHSs, and analyse the conditions under which a stable G–L coexistence occurs. Subsequently, we focus on the value of the second virial coefficient B_2 at the critical point. Finally, selected phase diagrams are compared with Monte Carlo simulations.

2.3.1 Phase Diagrams

In this Section, phase diagrams are presented for dispersions containing HCY interacting spheres plus PHSs. As expected from the pair potentials (see Fig. 1.1), the parameters that determine phase coexistence are the relative depletant concentration ϕ_d^R, the relative size of the depletant $q \equiv 2\delta/\sigma$; and the strength $\beta\epsilon$ and range q_Y of the HCY interaction. In Fig. 2.2, FVT phase diagrams are presented for mixtures of HSs ($\epsilon = 0$) and PHSs for a wide range of q-values. These provide a baseline for quantifying the effects of additional direct colloid–colloid interactions. In absence of

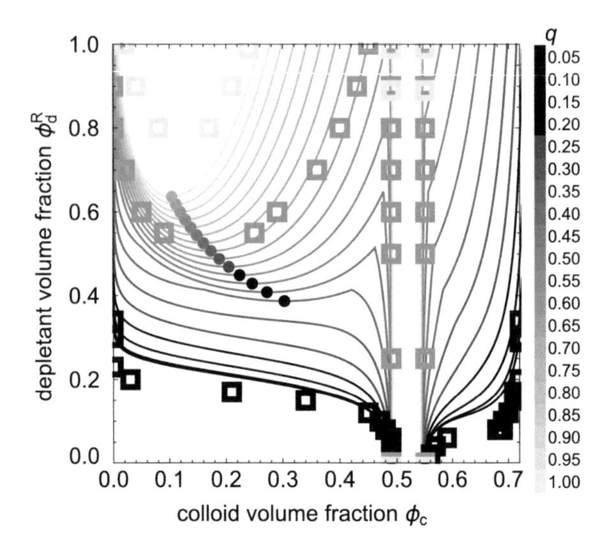

Fig. 2.2 Phase diagrams for mixtures of hard spheres (HSs) plus penetrable hard spheres (PHSs) for a wide range PHS–to–HS size ratios (q) between 0.05 and 1 as indicated. The cases $q = 0.1$, $q = 0.6$ and $q = 1$ are compared with independent simulation results (squares, taken from [6]). Filled circles refer to the critical points when G–L coexistence is thermodynamically stable ($q \gtrsim 0.33$)

depletants, F–S coexistence is found for $\phi_c \in \{0.49, 0.55\}$ as computed from the fluid and solid equations of state (see Chap. 1), in close agreement with computer simulations [18]. Colloidal G–L phase coexistence is obtained for ϕ_d^R values above the critical point and is characterised by a distinctive U-shape. For short-range depletion attractions ($q \lesssim 0.33$) only colloidal fluid–solid (F–S) coexistence is observed (the G–L coexistence is metastable). For long-range depletion attractions ($q \gtrsim 0.33$), stable colloidal G–L coexistence takes place. The transition between these two regimes is defined by the critical end point, CEP [1, 19]. Three binodal curves are compared with (independent) simulation results [6].

In Fig. 2.3a, phase diagrams calculated for a long-range Yukawa interaction ($q_Y = 1$) combined with (relatively) short-range depletion attraction ($q = 0.25$) are presented. For particles with repulsive HCY interactions (dashed curves) there is only F–S coexistence and the coexistence points are located at higher values of ϕ_d^R as the repulsions become stronger (increasing $|\beta\epsilon|$). There is G-L coexistence for this combination of $\{q_Y, q\}$ when the attractions are strong enough ($\beta\epsilon > 0.25$). Such G–L coexistence leads to a narrow liquid window for ($\beta\epsilon = 0.66$), and for stronger attractions the whole phase space gets unstable (demixing). The latter indicates that the thermodynamics of the CPM are dominated by the state of the depletant-free HCY suspension. While HCY repulsions hardly affect the F-S phase in $0.49 \lesssim \phi_c \lesssim 0.55$, additional HCY attractions broaden this F–S coexistence region. As expected, added direct repulsions between colloidal spheres stabilize (widen) the stability regions of colloid–polymer mixtures while direct attractions decrease the extent of the stable regions.

Phase diagrams obtained as a result of long-range depletion ($q = 1$) and short-range HCY interactions ($q_Y = 0.15$) are presented in Fig. 2.3b. Independently of the strength and nature of the HCY interaction, the characteristic U-shape of G-L coexistence is always present for the combination $\{q, q_Y\} = \{1, 0.15\}$. There

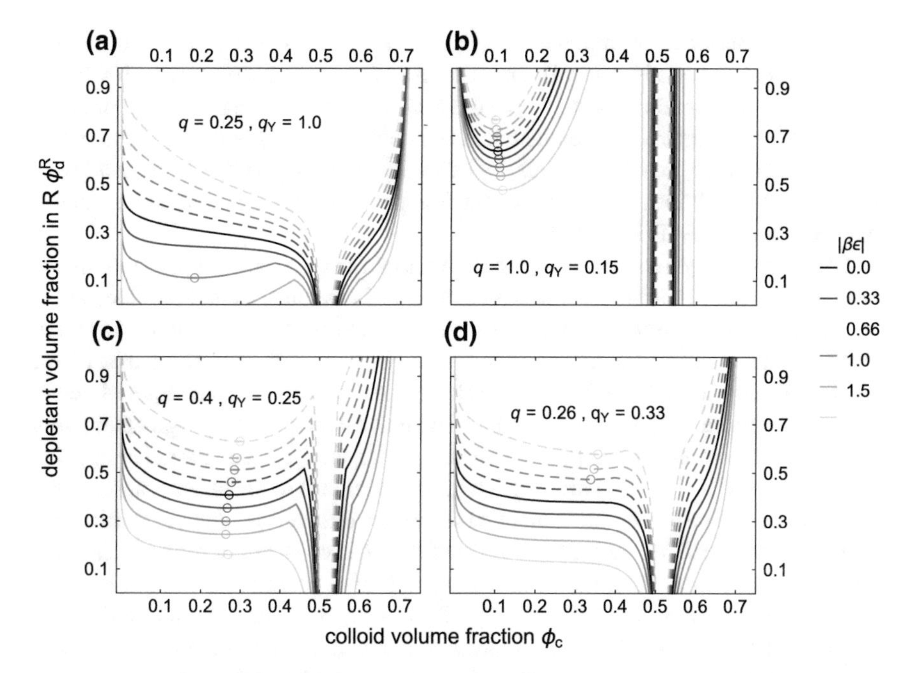

Fig. 2.3 Phase diagrams in the reservoir representation for a mixture of hard core Yukawa (HCY) spheres and PHS depletants for several relative depletant sizes q and ranges of HCY interaction q_Y with varying HCY strength ϵ as indicated. The G–L critical points are indicated by circles. Black curves hold for $\epsilon = 0$, corresponding to HS-PHS mixtures. Dashed curves correspond to HCY repulsions, while solid ones to attractions

is a remarkable increase of the width between the two vertical curves corresponding to F–S coexisting phase at low ϕ_d^R when direct attractions are incorporated (i.e., the HS F–S coexistence broadens due to added direct attractions). Even though the Yukawa attraction is short-ranged, the G–L binodals are clearly shifted. There is always G–L coexistence at $\phi_c < 0.4$ for these sets of $\{q_Y, q, \epsilon\}$ values. Contrary to the previous case, for this set of $\{q, q_Y\}$-values the G–L coexistence seems to follow what is expected when no additional direct colloid–colloid interactions are taken into account.

Different combinations of Yukawa and depletion interactions can be explored by systematically varying the system parameters $\{q, q_Y, \epsilon\}$. The most striking observation is that the direct Yukawa interactions change the topology of the phase diagrams with respect to the HS case, inducing a stable G–L coexistence when only F–S is present for HSs and *viceversa*. For instance, in Fig. 2.3c the G–L coexistence vanishes upon increasing the strength of the direct HCY (short-ranged) attraction (becomes metastable with increasing ϵ). In Fig. 2.3d binodals are presented for the set $\{q, q_Y\} = \{0.33, 0.25\}$, the corresponding critical end point (CEP) of purely attractive HCY fluids and HSs in a sea of PHSs (see next Section). The difference for HCY spheres in a sea of PHSs with respect to the HS case is small and follows the trends previously described. All binodals in Fig. 2.3d exhibit a shape of the fluid branch that is critical (or nearly critical), and a narrow G–L coexistence arises for repulsive Yukawa interactions above $|\beta\epsilon| = 0.33$. Due to combined short-ranged indirect attractions and short-ranged repulsions, G–L coexistence takes place, while it was metastable for HSs plus PHSs. Thus, by carefully analysing the CEP, one can reveal under which conditions a G–L coexistence appears or disappears in cases when it was present in a HS+PHS mixture.

The following general trends can be withdrawn from computed phase diagrams plotted in Fig. 2.3. Additional Yukawa attractions between colloidal particles lower the depletant concentration required for phase coexistence. This applies independently of the nature of the coexistence phase (G–L or F–S) and the ranges of the interactions considered. Equivalently, added repulsions increase the required depletant concentration for phase coexistence. The latter is in qualitative agreement with computer simulations and experimental results for charged nanoparticles in polymeric solutions [9, 11–13]. Not surprising, the difference in the phase diagram with respect to the HS cases is more dramatic for larger values of $|\beta\epsilon|$ and q_{Y}. Sufficiently strong, short-ranged Yukawa attractions turn the G–L coexistence region eventually metastable, inducing a thermodynamically stable F–S coexistence even at low colloid volume fractions. This shows that it is possible to modify the phase diagram topology of CPMs by tuning the Yukawa pair potential. Such tuning of the direct colloid interactions in experimental systems can be achieved for example by modifying the double layer [20] or brush repulsion [21, 22] or in near-critical solvent mixtures [23].

2.3.2 *Critical End Point*

The critical end point (CEP) is the main indicator of the topology of the phase diagrams, as it marks the limits at which a certain isostructural coexistence is stable [19]. Here, we focus on the CEP of the colloidal gas–liquid (G–L) coexistence. For a suspension of attractive HCY spheres, [24] the CEP is $q_{\mathrm{Y}}^{\mathrm{cep}} \approx 0.26$, while for HSs in a sea of PHSs, FVT provides [25] $q^{\mathrm{cep}} \approx 0.33$. These values establish a natural reference point for quantifying the effects of added Yukawa interactions, as hinted at in Fig. 2.3d. The calculated q-values at the CEP are summarised in Fig. 2.4 (right panel): if a chosen point in $\{q, q_{\mathrm{Y}}\}$ lies below the CEP curve at a given ϵ, the G–L coexistence is metastable. Above $\{q, q_{\mathrm{Y}}\}$, stable colloidal G–L coexistence appears. On the right panel, the minimum depletant concentration required for a stable G–L coexistence ($\phi_{\mathrm{d}}^{\mathrm{R,cep}}$) are also presented. As expected, $\phi_{\mathrm{d}}^{\mathrm{R,cep}}$ lowers with added Yukawa attractions and increases with Yukawa repulsions (see Fig. 2.4, left panel).

We have divided the ranges required for stable G–L coexistence in terms of the CEP-ranges of the HCY and depletion interactions into regions (I)–(IV), marked on the right panel of Fig. 2.4. (I): $q > q^{\mathrm{cep}}$, $q_{\mathrm{Y}} < q_{\mathrm{Y}}^{\mathrm{cep}}$. Due to a short-range direct HCY attraction that destabilizes colloidal G–L coexistence, the range of the depletion

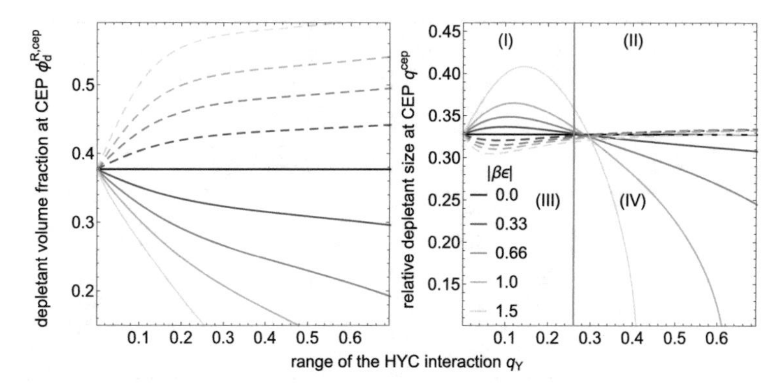

Fig. 2.4 Critical end point curves. Left panel: relative size of depletant at the CEP. Right panel: relative depletant concentration in the reservoir, $\phi_d^{R,cep}$, at the critical depletion interaction endpoint, q^{cep}, for various added HCY interactions as predicted by free volume theory. All curves coincide at $\epsilon = 0$, the hard sphere case (black line). On the left panel, the vertical line corresponds to the CEP of attractive HCY spheres. For the description of the regions (I)–(IV), see main text

interaction needs to be increased in order to still achieve a stable G–L coexistence, see the vanishing G–L binodals in Fig. 2.3c. Added short-range HCY repulsions do not destabilize the G-L coexistence. (II): $q > q^{cep}$, $q_Y > q_Y^{cep}$. Additional HCY repulsions barely affect the stability of the G–L coexistence [see Fig. 2.3c]. (III): $q < q^{cep}$, $q_Y < q_Y^{cep}$. There is no G–L coexistence present unless $q \approx q^{cep}$ and $q_Y \approx q_Y^{cep}$. In this (surprising) case, short-range direct colloid–colloid repulsions may induce a stable (narrow) G–L coexistence when it was not present in the HS case, as shown in Fig. 2.3(d). (IV): $q < q^{cep}$, $q_Y > q_Y^{cep}$. Sufficiently long-ranged HCY attractions may induce stable G–L coexistences when these were not present in the HS case, see Fig. 2.3a, for $\beta\epsilon = 0.66$. Note, however, that stable G–L coexistence does not necessarily imply the presence of a G–L critical point at finite depletant concentration when strong and long-ranged Yukawa and depletion attractions are combined, see Fig. 2.3a for strong attractions. As observed from the CEP behaviour, the system is governed either by the depletion or the HCY pure systems far from their relative CEPs. Let us take, for example, the HCY-attractive cases (a) and (b) in Fig. 2.3. In (a), the G–L of the present for HSs+PHSs may simple 'sink' into the F–S coexistence of the pure attractive HCY system when ϵ is large enough; in (b) the G–L coexistence appears because it was present in the HS+PHS mixture; it does not get metastable even at relatively high ϵ-values.

2.3.3 Second Virial Coefficient at the Critical Point

It is worthwhile analysing the value of the second virial (B_2) coefficient at the critical point (CP), given the tools for systematically calculating the CP from FVT. In Fig. 2.5 the normalised second osmotic virial coefficient ($B_2^* \equiv B_2/v_c$) is plotted as a function

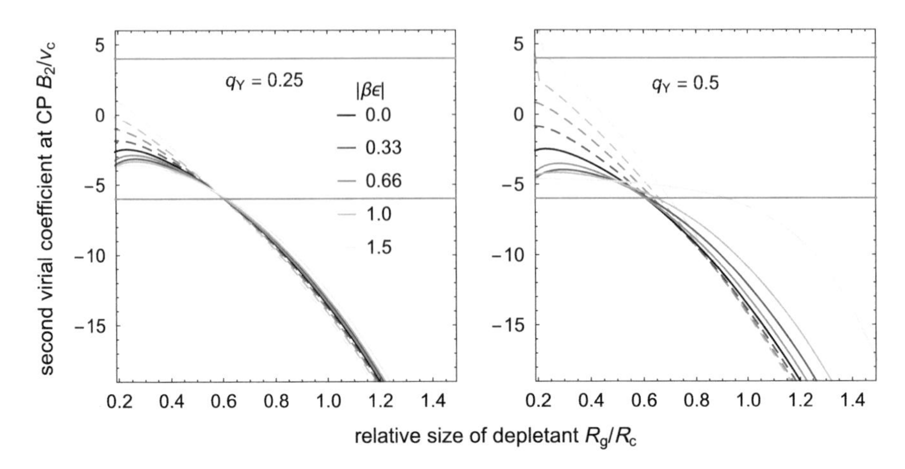

Fig. 2.5 Second virial coefficient at the G–L critical point with increasing relative size of depletant q for a collection of q_Y values for the HCY strengths ϵ indicated. Dashed curves correspond to added repulsive interactions, and solid ones to attractive ones. Horizontal lines correspond to the HS value $[B_2^*(HS) = 4]$ and the VL criterion $[B_2^*(\text{CP})_{\text{V-L}} = -6]$. The black solid line holds for the hard spheres plus depletants case

of q for two different q_Y-values (below and above the CEP) for a collection of ϵ-values. It is clear the Vliegenthart-Lekkerkerker (VL) criterion ($B_2^* = -6$) does not hold for mixtures HSs and PHSs, even when the added direct attractions are sufficiently long-ranged. This is due to the indirect nature of the depletion interaction: the VL criterion is based upon *direct* attractions [26]. For sufficiently long-ranged repulsive potentials B_2^* lies just below the HS value ($B_2 = 4v_c$) for small q-values. As expected, indirect attractions lead to different physical behaviour. As a consequence, when q is high enough, B_2^* at the critical point can be smaller than -6 even for additional strong, long-ranged direct attractions (as can be seen for instance for $q_Y = 0.5$ and $\beta\epsilon = 1.5$ in Fig. 2.5). Not surprisingly, low q_Y-values exhibit trends closer to the HS case than high ones.

2.3.4 Comparison with Monte Carlo Simulations

Finally, results from Monte Carlo simulations are presented which were conducted in order to verify the accuracy of the theoretical FVT predictions. In a simulation box at constant volume and temperature, a collection of $N_c = 256$ HSs, interacting via a depletion potential [W_{AOV}, Eq. (1.2)] *plus* a HCY potential [W_{HCY}, Eq. (1.1)], was employed (NVT ensemble). A collection of $\{\phi_c, \phi_d^R\}$ state points in the phase diagram is considered for each set of $\{q, q_Y, \beta\epsilon\}$. The free energy of the ensemble at each $\{\phi_c, \phi_d^R\}$ state point is estimated using the λ-integration method for the depletion interaction, with a ten-point Gauss–Legendre integration (following [2, 9, 27] and

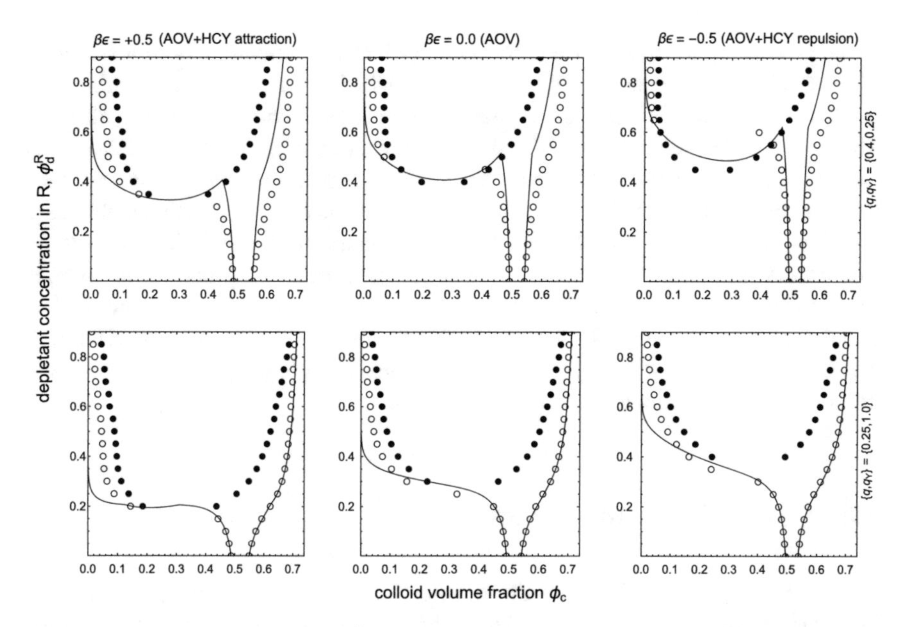

Fig. 2.6 Phase diagrams in the $\{\phi_c, \phi_d^R\}$ phase space for hard-core Yukawa spheres plus penetrable hard spheres as predicted via MC simulations for the $\{q, q_Y, \beta\epsilon\}$ system parameters considered. Circles correspond to F-S coexistence, whereas discs indicate G–L coexistences

references therein). The MC simulations ran for 32000 lattice sweeps at each λ value, enough for the system to equilibrate in the state points studied. The considered free energy of the system was averaged over the last 3200 MC cycles at each state point. The collected energy can be associated with the (dimensionless) free energy of the system:

$$\widetilde{F}_{\mathrm{NVT}}^{\mathrm{MC}} = \left(\underbrace{\langle \sum_{i<j}^{N_c} W_{\mathrm{HCY}}(x_{\mathrm{ij}}) \rangle}_{\text{HCY contri.}} + \underbrace{\int_0^1 \mathrm{d}\lambda \sum_{i<j}^{N_c} \langle W_{\mathrm{AOV}}(x_{\mathrm{ij}}) \rangle}_{\lambda\text{-inte. depletion}} \right) \frac{v_c}{V_{\mathrm{MC}}} \frac{1}{k_B T} \approx \underbrace{\widetilde{F}_{\mathrm{HCY}}}_{\text{Analytical}} + \underbrace{\widetilde{F}_{\mathrm{AOV}}}_{\text{Fit}}, \tag{2.5}$$

where the angular brackets (⟨...⟩) indicate ensemble-averages. Even though already verified to be accurate, [24] we checked the feasibility of the FMSA. The match between the MC-averaged free energies and the analytical FMSA expressions enables to perform the λ-integration only over the depletion pair potential.

Phase diagrams obtained using the free energies obtained from MC simulations for HCY spheres plus the considered depletion potential are presented as the data points in Fig. 2.6. The trends observed match the ones of FVT for HCY spheres plus PHSs (solid curves). The F–S coexistence arising from the HS case is broader at high colloidal packing fraction than predicted by FVT at sufficiently high

$\{q, \phi_d^R\}$-values. This (most likely) reflects to the lack of accounting for multi-overlap of the depletion zones in the MC simulations, which is accounted for within FVT. In all cases studied, MC-predicted phase diagrams quantitatively match with FVT. Moreover, the results for HSs+PHSs slightly differ from the original calculations, [2] where no G–L coexistence was found for $q = 0.4$. Experimentally, G–L coexistence has been reported for $q > 0.3$ [1]. Future improvements of the simulation method are possible, mainly accounting for the multi-body nature of the depletion interaction, [6] without presuming the FMSA free energy (i.e., performing two λ-integrations for the free energy) or using binary mixtures of HSs and PHSs in a canonical or grand-canonical ensemble. This would enable a more precise comparison of our theory. Overall, there is good agreement in the trends of the results with respect to the influence of the additional Yukawa interactions.

2.4 Conclusions and Outlook

The quantitative match between the Monte Carlo-generated phase diagrams and those arising from free volume theory implies that the tuning knobs that determine the phase diagram of hard-core Yukawa (HCY) spheres in a sea of penetrable hard spheres (PHSs) have been identified correctly within our theoretical framework. The depletant concentration and relative depletant size plus the range and strength of the direct interactions allow controlling the stable phase regions of colloid–polymer mixtures. Deviation from the hard sphere (HS) case takes place as the direct HCY interactions become stronger, inducing different phase coexistence regions than those present for a pure suspension of HSs in a sea of polymeric depletants. Here we have identified how the colloidal gas–liquid critical end point, that determines the coexisting phases at a particular set of interaction potentials, is shifted due to combined interactions.

References

1. H.N.W. Lekkerkerker, R. Tuinier, *Colloids and the Depletion Interaction* (Springer, Heidelberg, 2011)
2. M. Dijkstra, J.M. Brader, R. Evans, J. Phys.: Condens. Matter **11**, 10079 (1999)
3. D.J. Ashton, N.B. Wilding, *Phys. Rev. E*, 2014, p. 031301. https://doi.org/10.1103/PhysRevE.89.031301
4. M. Majka, P.F. Góra, Phys. Rev. E **90**, 032303 (2014). https://doi.org/10.1103/PhysRevE.90.032303
5. L. Rovigatti, N. Gnan, A. Parola, E. Zaccarelli, Soft Matter **11**, 692 (2014). https://pubs.rsc.org/en/content/articlelanding/2015/sm/c4sm02218a/unauth#!divAbstract
6. M. Dijkstra, R. van Roij, R. Roth, A. Fortini, Phys. Rev. E **73**, 041404 (2006). https://journals.aps.org/pre/abstract/10.1103/PhysRevE.73.041404
7. J. Jover, A. Galindo, G. Jackson, E.A. Müller, A.J. Haslam, Mol. Phys. **113**, 2608 (2015). https://doi.org/10.1080/00268976.2015.1047425

8. A. Santos, M. López de Haro, G. Fiumara, F. Saija, J. Chem. Phys. **142**, 224903 (2015). https://doi.org/10.1063/1.4922031
9. A. Fortini, M. Dijkstra, R. Tuinier, J. Phys.: Condens. Matter **17**, 7783 (2005). http://stacks.iop.org/0953-8984/17/i=50/a=002
10. A.R. Denton, M. Schmidt, J. Chem. Phys. **122**, 244911 (2005). https://doi.org/10.1063/1.1940055
11. C. Gögelein, R. Tuinier, Eur. Phys. J. E **27**, 171 (2008). https://doi.org/10.1140/epje/i2008-10367-6
12. K. van Gruijthuijsen, R. Tuinier, J.M. Brader, A. Stradner, Soft Matter **9**, 9977 (2013). https://doi.org/10.1039/c3sm51432c
13. G. Pandav, V. Pryamitsyn, V. Ganesan, Langmuir **31**, 12328 (2015). https://doi.org/10.1021/acs.langmuir.5b02885
14. E. Dickinson, Food Hydrocoll. **52**, 497 (2016). https://doi.org/10.1016/j.foodhyd.2015.07.029
15. Y. Tang, B.C. Lu, J. Chem. Phys. **99**, 9828 (1993). https://aip.scitation.org/doi/10.1063/1.465465
16. A. Vrij, Pure Appl. Chem. **48**, 471 (1976). https://doi.org/10.1351/pac197648040471
17. H.N.W. Lekkerkerker, W.C.K. Poon, P.N. Pusey, A. Stroobants, P.B. Warren, Europhys. Lett. **20**, 559 (1992). https://doi.org/10.1209/0295-5075/20/6/015
18. W.G. Hoover, F.H. Ree, J. Chem. Phys. **49**, 3609 (1968). https://aip.scitation.org/doi/10.1063/1.1670641
19. C.F. Tejero, A. Daanoun, H.N.W. Lekkerkerker, M. Baus, Phys. Rev. Lett. **73**, 752 (1994). https://journals.aps.org/prl/abstract/10.1103/PhysRevLett.73.752
20. C.P. Royall, M.E. Leunissen, A. van Blaaderen, J. Phys.: Condens. Matter **15**, S3581 (2003). http://stacks.iop.org/0953-8984/15/i=48/a=017
21. P. Uhlmann, H. Merlitz, J.-U. Sommer, M. Stamm, Macromol. Rapid Commun. **30**, 732 (2009). https://doi.org/10.1002/marc.200900113
22. F. Lo Verso, L. Yelash, S.A. Egorov, K. Binder, Soft Matter **8**, 4185 (2012) https://doi.org/10.1039/c2sm06836b
23. M.T. Dang, A.V. Verde, V.D. Nguyen, P.G. Bolhuis, P. Schall, J. Chem. Phys. **139**, 094903 (2013). https://doi.org/10.1063/1.4819896
24. R. Tuinier, G.J. Fleer, J. Phys. Chem. B **110**, 20540 (2006). https://pubs.acs.org/doi/abs/10.1021/jp063650j
25. G.J. Fleer, R. Tuinier, Adv. Colloid Interface Sci. **143**, 1 (2008) https://doi.org/10.1016/j.cis.2008.07.001
26. R. Tuinier, M.S. Feenstra, Langmuir **30**, 13121 (2014). https://doi.org/10.1021/la5023856
27. M. Dijkstra, Phys. Rev. E **66**, 021402 (2002). https://doi.org/10.1103/PhysRevE.66.021402

Chapter 3
Depletion-Driven Solid–Solid Coexistence in Colloid–Polymer Mixtures

3.1 Introduction

We focus in this chapter on a model system for the depletion attraction, which arises when hard spheres (HSs, with radius R) are mixed with penetrable hard spheres as depletants (PHSs, with radius δ). PHSs can freely overlap with each other, but feature hard-core repulsions with HSs [1]. All length scales involved in the problem of interest are captured in terms of the depletant-to-colloid size ratio $q \equiv 2\delta/\sigma$. PHSs are an approximation to polymeric depletants at relatively low concentrations in θ-solvents [2].

Different thermodynamic approaches can be followed to reveal the phase behaviour of these model colloid–polymer mixtures (CPMs). For instance, depletion effects may be mapped onto an effective pair potential between the colloidal particles, as done originally by Asakura and Oosawa [3, 4] and Vrij [1], often denoted as the AOV potential (see Chap. 1). These pair potentials can be used in Monte Carlo routines or other theoretical approaches for phase stability studies [5, 6]. More sophisticated approaches account for multiple overlap of depletion zones [7] or the statistics of polymer chains [8, 9].

Independent of the approach followed, it is the size ratio q that defines the possible thermodynamically stable phases. It is well-known that a collection of HSs only exhibits a fluid–solid (F–S) phase transition [10]. Upon adding large depletants ($q \gtrsim 0.33$), the range of the effective attractions is sufficient to display additionally an isostructural F_1–F_2 equilibrium: the colloidal gas–liquid (G–L) transition [11–13]. In this chapter we show that, in the opposite limit of sufficiently small PHSs, an isostructural solid–solid phase transition can be realised on purely entropic grounds, without invoking explicit pair potentials.

© Springer Nature Switzerland AG 2019

Á. González García, *Polymer-Mediated Phase Stability of Colloids*,
Springer Theses, https://doi.org/10.1007/978-3-030-33683-7_3

3.2 Geometrical Free Volume Fraction in the Solid State

We study the model CPM on the basis of free volume theory (FVT), developed by Lekkerkerker et al. in the early 1990s [14, 15]. FVT is an instructive approach to describe the phase behaviour of CPMs, because it accounts explicitly for the partitioning of depletants over the different phases by considering the free (accessible) volume of the depletants in the HS dispersion [11]. Several improvements have been incorporated into FVT to bring it closer to reality [16–18], for instance, by accounting for configurations of large polymer chains *around* smaller colloidal particles [2, 19], that effectively increase the free volume available to the depletants.

One remaining particularity of FVT for HS–PHS mixtures in the small depletant limit ($q \lesssim 0.1$) is the prediction [11] that the F–S binodal becomes *independent* of q, see Fig. 3.1 (grey curve). This is unphysical, because in the small q limit the depth ϵ of the depletion potential for two spheres at contact in the AOV model is [11] (see Chap. 1)

$$\frac{W_{\mathrm{AOV}}(r=\sigma)}{k_B T} \simeq -\frac{3}{2}\frac{\phi_{\mathrm{d}}^{\mathrm{R}}}{q}, \tag{3.1}$$

where $k_B T$ is the thermal energy and $\phi_{\mathrm{d}}^{\mathrm{R}}$ the bulk volume fraction of PHSs. If the location of the binodal in terms of depletant concentration would be independent of q, the potential depth $W_{\mathrm{AOV}}(r=\sigma)$ at the binodal would diverge. Instead, one would expect $W_{\mathrm{AOV}}(r=\sigma)$ at the binodal to be nearly independent of depletant size, so that the depletant concentration at the binodal should be approximately proportional to q. Experiments indeed show this trend (Fig. 3.1, data points). We show that this unphysical behaviour of FVT for small q is a consequence of an imprecise description of the free volume fraction for PHSs in the HS solid phase: an improved description brings FVT in line with experimental results and, crucially, yields the aforementioned solid–solid phase coexistence.

The core idea of the FVT used was introduced in Chap. 1. For HSs mixed with PHSs the dimensionless grand potential $\widetilde{\Omega}_k \equiv \frac{\Omega_k v_{\mathrm{c}}}{V k_B T}$ (with v_{c} the HS volume and V the system volume) has the form:

$$\widetilde{\Omega}_k = \widetilde{F}_k - q^{-3}\alpha_k \widetilde{\Pi}_{\mathrm{d}}^{\mathrm{R}}, \tag{3.2}$$

where $\widetilde{F}_k$ is the dimensionless free energy of pure HSs and the second term on the right hand side incorporates depletion effects (see Chap. 1, where also the expressions for the free energies $\widetilde{F}_k$ of a HS system are presented). Implicitly, it is assumed that the depletant concentration does not affect α. For colloidal hard spheres mixed with PHS depletants, α reads:

$$\alpha_{\mathrm{F}} = \underbrace{(1-\phi_{\mathrm{c}})}_{\text{point depl.}} \underbrace{\exp(-Q_{\mathrm{s}})}_{\text{small depl.}} \underbrace{\exp(-q^3 \widetilde{\Pi}_{\mathrm{F}}^{\mathrm{o}})}_{\text{cavity}}, \tag{3.3}$$

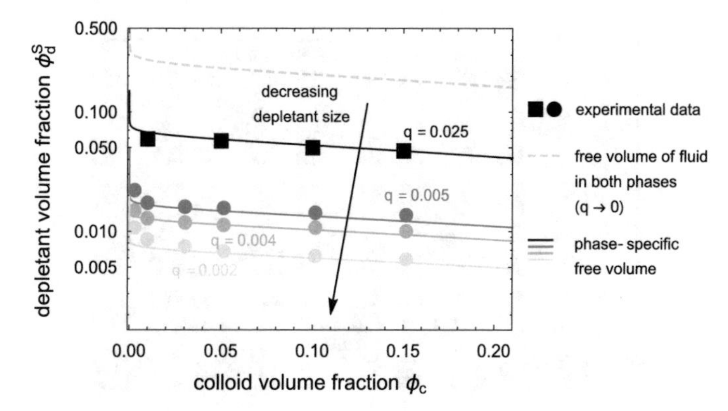

Fig. 3.1 Fluid branch of the fluid–solid binodals from free volume theory, (dashed curve) using free volume expressions of the fluid phase for both phases in the limit of depletant-to-colloid size ratio $q \to 0$; and (black and grey curves) using phase-specific free volume expressions for the indicated q-values. Symbols represent experimental results from (squares) Piazza and Di Pietro [20] and (circles) Bibette et al. [21]

with

$$Q_s = 3qy + \frac{y}{2}q^2[6 + 9y], \tag{3.4}$$

and

$$y = \frac{\phi_c}{1 - \phi_c}. \tag{3.5}$$

Traditionally, Eq. (3.3) is used not only for the fluid phase, but also for the solid state [11]. The argument for this is that α in the solid phase is very low, so that the absolute difference between α_F and the actual free volume of a solid phase will be small. A slightly improved estimate for the free volume in a solid phase can be obtained by replacing $\widetilde{\Pi}_F^o$ in Eq. (3.3) by the osmotic pressure of a HS solid, $\widetilde{\Pi}_S^o$. However, this does not drastically improve the phase diagrams in the small q limit, as α is dominated by the first two terms in Eq. (3.3), due to the q^3 dependence of the third term. This motivated us to investigate α in the solid state in more detail.

We determine the free volume fraction α_S of the solid state on the basis of geometrical arguments. A system in a solid phase state can be defined via its unit cell (UC). Hence, the volume of the system V is directly the unit cell volume $V = V_{UC}$. In a solid state of static HSs where depletants are present, three different regions can be identified: there is no overlap of depletion zones, there is overlap of depletion zones, and there is no free volume for depletants. An illustrative scheme of the method set is presented in Fig. 3.2.

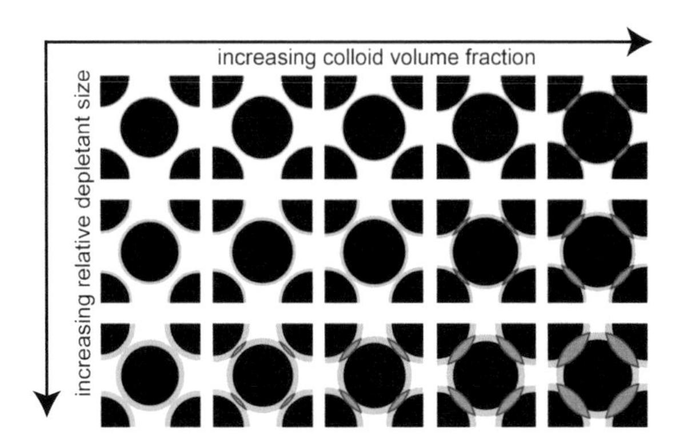

Fig. 3.2 $(1, 0, 0)$ plane representation of the unit cell of an FCC lattice with increasing colloid volume fraction and relative depletant size. In black, the colloidal hard spheres. Light grey stands for the depletion zones. Dark grey regions represent the overlap of the depletion zones. White space within each unit cell corresponds to the free volume for depletants

In general, the free volume fraction in a solid state can be written by looking at the unit cells as:

$$\alpha_S = \begin{cases} 1 - \frac{N_c v_{dep}}{V_{UC}} & \text{for } \tilde{r} < 1+q, \\ 1 - \left[\frac{N_c v_{dep}}{V_{UC}} - \frac{\kappa v_{overl}}{V_{UC}}\right] & \text{for } \tilde{r} > 1+q, \\ 0 & \text{if no free volume,} \end{cases} \tag{3.6}$$

where $\tilde{r} \equiv r/\sigma$ is the distance between nearest neighbours within the UC (which depends on ϕ_c), v_{dep} is the volume of a depletion zone, N_c is the actual number of colloidal particles within the UC, and κ is the total number of depletion zone overlaps within the UC. For the FCC lattice:

$$\tilde{r} = \left(\phi_c^{cp}/\phi_c\right)^{1/3}, V_{UC} = 2^{3/2} r^3, N_c = 4, \kappa = 24 . \tag{3.7}$$

Note that r is obtained via *geometrical* arguments: the structure of the UC does not depend on the colloid concentration. The term v_{dep} reads:

$$v_{dep} = v_c(1+q)^3. \tag{3.8}$$

On the other hand, the volume of overlap of depletion zones is [11]:

$$v_{overl} = v_c \left[1 + q - \left(\frac{\phi_c^{cp}}{\phi_c}\right)^{\frac{1}{3}}\right]^2 \left[1 + q + \frac{1}{2}\left(\frac{\phi_c^{cp}}{\phi_c}\right)^{\frac{1}{3}}\right]. \tag{3.9}$$

The colloid volume fraction at which overlap of depletion zones just starts to take place (ϕ_c^*) follows as:

$$\tilde{r} = 1 + q, \quad \phi_c^* = \phi_c^{cp}/\tilde{v}_{exc}^{o}, \tag{3.10}$$

with

$$\tilde{v}_{exc}^{o} = v_{dep}/v_c = (1+q)^3. \tag{3.11}$$

Finally, the colloid concentration at which there is no free volume for depletants follows from the condition at which

$$\sqrt{2}\tilde{r} > (q+1), \quad \phi_c > 2^{3/2}\phi_c^* . \tag{3.12}$$

With all the ingredients at hand, it follows from Eq. (3.6) that

$$\alpha_S = \begin{cases} 1 - \phi_c \tilde{v}_{exc}^{o} & \text{for } \phi_c < \phi_c^* \text{ (no overlap)}, \\ 1 - \phi_c \tilde{v}_{exc}^{*} & \text{for } \phi_c^* \leq \phi_c < 2^{3/2}\phi_c^* \text{ (overlap)}, \\ 0 & \text{otherwise}, \end{cases} \tag{3.13}$$

with

$$\tilde{v}_{exc}^{*} \equiv \tilde{v}_{exc}^{o} - 6v_{overl}/v_c. \tag{3.14}$$

Monte Carlo (MC) simulations provide a way to verify the validity of Eq. (3.13). To this end we performed simulations on an NVT-ensemble of $N_c = 256$ HSs with an initial FCC state of 2×10^5 MC steps (one step corresponding to N_c trials of moving a randomly selected particle). In each of the last 50% of the MC steps, a virtual attempt was made to insert a PHS. The free volume fraction is given as the number of accepted insertions over the number of trials performed. The reported values correspond to the averages of 10 different runs.

The free volume fractions for PHSs in the HS solid state are presented in Fig. 3.3 as obtained from Eqs. (3.3) to (3.13) and are compared to our MC simulations. We focus first on the case of $q = 0.2$. Up to $\phi_c \approx 0.65$, the free volume from MC simulations follows the SPT fluid prediction, Eq. (3.3). For larger HS volume fractions, there is a crossover to our new prediction for the solid state, Eq. (3.13). For $q = 0.1$, this crossover happens also at $\phi_c \approx 0.65$. For even smaller depletants ($q = 0.05$), both the fluid and the solid state free volume fractions become an adequate description at lower volume fractions, but only the solid phase prediction matches the MC simulations at larger volume fractions. On each occasion, in the close packed regime, $\phi_c \to 0.74$, Eq. (3.13) predicts the MC results much better than Eq. (3.3). Thus, α_F *underestimates* the free volume for depletants for $q \lesssim 0.2$ at high ϕ_c (see Fig. 3.3). It is important to note that the thermodynamic properties of colloid–polymer mixtures do not only depend on the absolute value of α, but also on its derivative. As can be seen from Fig. 3.3 (left panel), the slopes near close packing are also different and this bears important consequences for the phase behaviour.

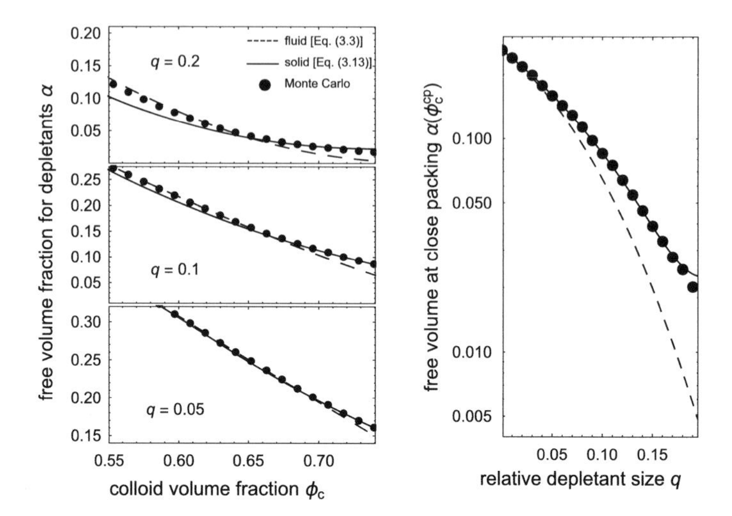

Fig. 3.3 Comparison between the free volume fractions for penetrable hard spheres in a system of hard spheres from scaled particle theory (dashed curves) for the fluid phase [Eq. (3.3)], from our analytical expression (solid curves) for the solid phase [Eq. (3.13)], and from Monte Carlo simulations (filled circles)

3.3 Phase Diagrams and Solid–Solid Critical End Point

We now turn our attention to the phase behaviour of these HS–PHS mixtures. Calculation of phase diagrams is straightforward from Eq. (3.2) and basic thermodynamics (see Chap. 1). Figure 3.4 shows phase diagrams for $q \leq 0.1$ using our expression [Eq. (3.13)] for the free volume of the solid phase (black curves) and following the usual practice of using the free volume of the fluid phase [Eq. (3.3)] for the solid phase (solid grey curves). The pure HS F–S coexistence is recovered at $\phi_{\rm d}^{\rm R} = 0$ [10]. Contrary to the phase diagrams generated when the fluid free volume $\alpha_{\rm F}$ is used for the solid phase, we now do observe a shift of the binodal to lower depletant concentrations upon decreasing q, in agreement with experiments (see also Fig. 3.1, curves through the experimental points). Further, we observe a shift of the F–S coexistence towards higher HS volume fractions with increasing $\phi_{\rm d}^{\rm R}$. At small $\phi_{\rm d}^{\rm R}$ the F–S coexistences quantitatively follow the predictions from thermodynamic perturbation theory [22] (dashed grey lines, see Sect.3.A).

The phase diagrams for $q \lesssim 0.1$ also display an *isostructural solid–solid* (S_1–S_2) coexistence. Due to the piecewise nature of $\alpha_{\rm S}$ [see Eq. (3.13)], the curvature of the S_1–S_2 binodal to the left and right of the critical point differs. This S_1–S_2 critical point is found precisely at the HS volume fraction $\phi_{\rm c}^*$ where overlap of depletion zones starts to take place: the colloid–polymer mixture phase separates into a high density solid with large overlap of depletion zones and a dilute solid with no overlap. We note here that the (commonly applied) second derivative condition of a critical point (see Chap. 1) depends on the nature of the functions involved, and is not a

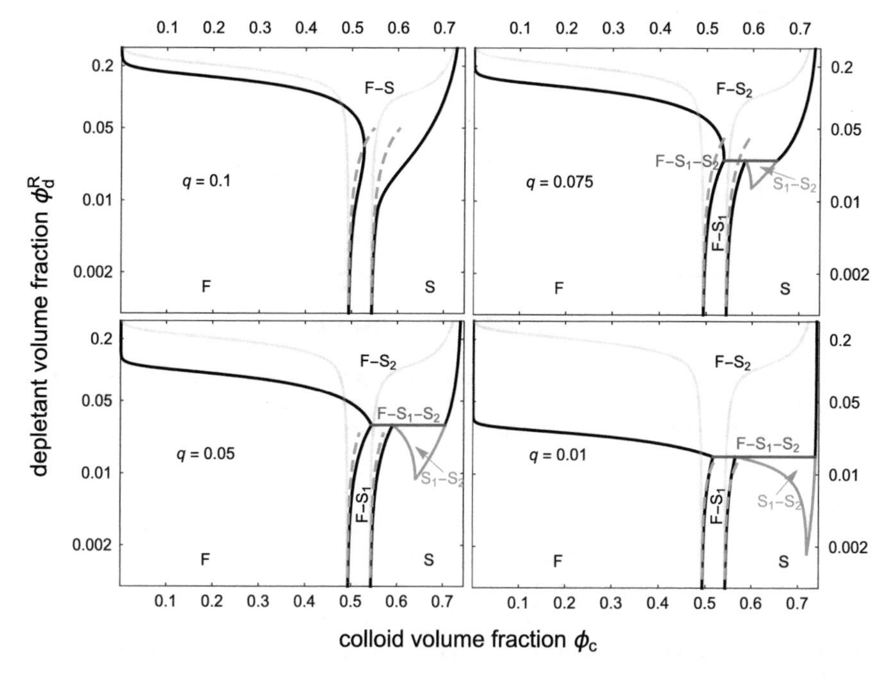

Fig. 3.4 Phase diagrams of hard spheres mixed with penetrable hard spheres as depletants for various relative depletant sizes q. The various curves are obtained using the solid free volume [Eq. (3.13)] for the solid phase (solid curves) and using the fluid free volume [Eq. (3.3)] for the solid phase (light grey curves). The triple lines and isostructural solid–solid binodals are indicated. The dashed grey curves correspond to the fluid–solid coexistence calculated using thermodynamic perturbation theory

necessary condition for defining a critical point [23, 24]. In fact, due to the different behaviour of α from the right and from the left of this CP, in this case the solid–solid critical point does not satisfy this condition.

Isostructural S_1–S_2 coexistence has been predicted theoretically [25–30] and reported in MC simulation studies [31, 32] for systems with explicit short-range attractive pair potentials, as well as for additive binary hard sphere mixtures [33]. Many of these authors hypothesized that mixing HSs with depletants might be a way to realise such an S_1–S_2 coexistence experimentally. So far, no experiments have demonstrated this transition. However, as in experiments, our computations do not invoke explicit pair potentials: the isostructural solid–solid coexistence arises from the partitioning (or entropy gain) of depletants. It must be noted that the wide presence of glassy states at such small ranges of attraction may make this isostructural solid–solid coexistence hard to find experimentally [34–36].

We now turn our attention to the critical behaviour of the S_1–S_2 coexistence, detailed in Fig. 3.5. We observe that the solid–solid critical point (CP) moves to lower HS volume fraction with increasing q, because $\phi_c^* \simeq \phi_c^{cp}(1 - 3q)$ for small q [see Eq. (3.13)]. This trend is in qualitative agreement with results from Dijkstra

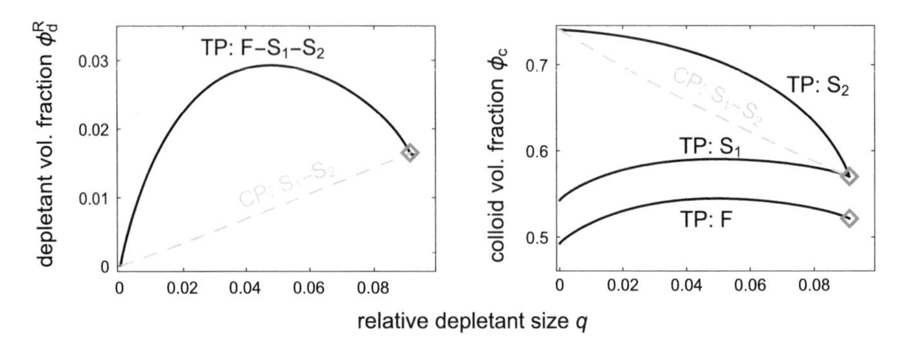

Fig. 3.5 Depletant volume fraction (ϕ_d^R) and colloid volume fraction (ϕ_c) at the fluid–solid–solid triple point (black curves) and solid–solid critical point (grey dashed curves) as a function of the relative depletant size q. The open symbols denote the critical endpoint

et al. for binary mixtures of hard spheres of different sizes [33]. Similarly, the PHS volume fraction at the CP increases linearly with q, since the depth of the potential [Eq. (3.1)] at the CP is expected to be approximately constant. Indeed, we find a limiting value of $W(r = \sigma) \approx -0.25k_BT$ for $q \to 0$ from the ratio of q and ϕ_d^R at the CP in terms of the AOV potential. This is considerably lower than found previously for HSs interacting through square-well [31] and Yukawa potentials [32], for which limiting contact potentials at the CP of $W(r = \sigma) \approx -0.59k_BT$ and $-1.49k_BT$ were found, respectively. This suggests that the S$_1$–S$_2$ transition is sensitive to the precise shape of the (effective) interactions. In particular, we expect the shape of the binodal near the S$_1$–S$_2$ critical point to become more rounded in real systems, as Brownian motion makes the transition from non-overlapping to overlapping depletion zones more gradual. For the triple point (TP), a soft re-entrant behaviour is observed: the PHS volume fraction first increases and then decreases, in agreement with previous studies on hard spheres with short-range Yukawa attractions [30].

The conditions for which a critical point becomes metastable is marked by the critical endpoint (CEP). More specifically, a CEP is found when two phases are at a CP while they are in equilibrium with a third distinct phase, for instance when the S$_1$–S$_2$ CP coexists with the fluid phase. This is identified graphically in Fig. 3.5, where the S$_1$–S$_2$ CP meets the TP. Thus, we find an upper q-limit for solid–solid coexistence of about $q \approx 0.091$; for larger PHSs the S$_1$–S$_2$ coexistence becomes metastable.

3.4 Conclusion

We have shown that mixtures of hard spheres plus small penetrable hard spheres as depletants exhibit an isostructural *solid–solid* phase coexistence. The finding of this phase coexistence follows from an accurate description of the free volume for depletants in the HS solid phase within free volume theory. Crucially, the solid–solid coexistence is caused solely by the partitioning of the depletants—without invoking explicit pair potentials—reflecting the situation of depletion-induced phase separation in experiments. The solid–solid coexistence is terminated by a critical endpoint at a depletant-to-colloid size ratio of $q \approx 0.091$. Our work contributes to the fundamental understanding of phase transitions in mixed colloidal dispersions and paves the way towards experimental realization of this solid–solid coexistence.

3.A TPT for the AOV Potential

In this short section we describe the approach followed to justify the shift of the fluid–solid (F–S) coexistence towards higher packing fractions upon addition of small depletants. Following standard thermodynamic perturbation theories [22] (previously applied to highly-screened repulsive interactions [37]), we consider an effective sphere of interaction whose diameter σ' is calculated via

$$\sigma'/\sigma = 1 + \int_1^\infty \mathrm{d}\tilde{r}\,(1 - \exp[-\beta W_{\mathrm{AOV}}(\tilde{r})])\,, \tag{3.15}$$

with the Asakura–Oosawa–Vrij (AOV) depletion pair potential given as in Eq. (1.2). The integral given in Eq. (3.15) can be solved numerically for all $\{q, \phi_{\mathrm{d}}^{\mathrm{R}}\}$, providing σ'. We then map the thermodynamic functions of a pure HS suspension with an effective packing fraction:

$$\phi_{\mathrm{c}}' = (\sigma'/\sigma)^3 \phi_{\mathrm{c}}.$$

By substituting $\phi_{\mathrm{c}} \leftrightarrow \phi_{\mathrm{c}}'$ on all canonical expressions, calculation of the F–S binodal is straightforward.

References

1. A. Vrij, Pure Appl. Chem. **48**, 471 (1976). https://doi.org/10.1351/pac197648040471
2. G.J. Fleer, R. Tuinier, Adv. Colloid Interface Sci. **143**, 1 (2008). https://doi.org/10.1016/j.cis.2008.07.001
3. S. Asakura, F. Oosawa, J. Chem. Phys. **22**, 1255 (1954). https://doi.org/10.1063/1.1740347
4. S. Asakura, F. Oosawa, J. Polym. Sci. **33**, 183 (1958). https://doi.org/10.1002/pol.1958.1203312618
5. A. Fortini, M. Dijkstra, R. Tuinier, J. Phys.: Condens. Matter **17**, 7783 (2005), http://stacks.iop.org/0953-8984/17/i=50/a=002
6. R. Roth, J. Phys.: Condens. Matter **22**, 063102 (2010), http://stacks.iop.org/0953-8984/22/i=6/a=063102?key=crossref.e8ded4d3b4b40a58361c21e1fbb1abc0
7. M. Dijkstra, R. van Roij, R. Roth, A. Fortini, Phys. Rev. E **73**, 041404 (2006), https://journals.aps.org/pre/abstract/10.1103/PhysRevE.73.041404

8. E.J. Meijer, D. Frenkel, J. Chem. Phys. **100**, 6873 (1994), https://www.google.com/search?client=firefox-b-ab&q=olloids+dispersed+in+polymer+solutions.+A+computer+simulation+study
9. J. Jover, A. Galindo, G. Jackson, E.A. Müller, A.J. Haslam, Mol. Phys. **113**, 2608 (2015). https://doi.org/10.1080/00268976.2015.1047425
10. W.G. Hoover, F.H. Ree, J. Chem. Phys. **49**, 3609 (1968), https://aip.scitation.org/doi/10.1063/1.1670641
11. H.N.W. Lekkerkerker, R. Tuinier, *Colloids and the Depletion Interaction* (Springer, Heidelberg, 2011)
12. M. Dijkstra, J.M. Brader, R. Evans, J. Phys.: Condens. Matter **11**, 10079 (1999)
13. F.L. Calderon, J. Bibette, J. Biais, Europhys. Lett. **23**, 653 (1993), http://stacks.iop.org/0295-5075/23/i=9/a=006
14. H.N.W. Lekkerkerker, Colloids Surf. **51**, 419 (1990), https://www.sciencedirect.com/science/article/abs/pii/016666229080156X
15. H.N.W. Lekkerkerker, W.C.K. Poon, P.N. Pusey, A. Stroobants, P.B. Warren, Europhys. Lett. **20**, 559 (1992). https://doi.org/10.1209/0295-5075/20/6/015
16. P.G. Bolhuis, E.J. Meijer, A.A. Louis, Phys. Rev. Lett. **90**, 068304 (2003). https://doi.org/10.1103/PhysRevLett.90.068304
17. A. Moncho-Jordá, A.A. Louis, P.G. Bolhuis, R. Roth, J. Phys.: Condens. Matter **15**, S3429 (2003), http://stacks.iop.org/0953-8984/15/i=48/a=004
18. K.J. Mutch, J.S. van Duijneveldt, J. Eastoe, Soft Matter **3**, 155 (2007). https://doi.org/10.1039/B611137H
19. P.G. Bolhuis, A.A. Louis, J.P. Hansen, E.J. Meijer, J. Chem. Phys. **114**, 4296 (2001). https://doi.org/10.1063/1.1344606
20. R. Piazza, G.D. Pietro, Europhys. Lett. **28**, 445 (1994), http://stacks.iop.org/0295-5075/28/i=6/a=012
21. J. Bibette, D. Roux, F. Nallet, Phys. Rev. Lett. **65**, 2470 (1990). https://doi.org/10.1103/PhysRevLett.65.2470
22. J.A. Barker, D. Henderson, J. Chem. Phys. **47**, 4714 (1967). https://doi.org/10.1063/1.1701689
23. P. Ehrenfest, Commun. Phys. Lab. Univ. Leiden **75b** (1933), https://www.lorentz.leidenuniv.nl/IL-publications/Ehrenfest.html
24. G. Jaeger, Arch. Hist. Exact Sci. **53**, 51 (1998), https://link.springer.com/article/10.1007/s004070050021
25. C.F. Tejero, A. Daanoun, H.N.W. Lekkerkerker, M. Baus, Phys. Rev. Lett. **73**, 752 (1994), https://journals.aps.org/prl/abstract/10.1103/PhysRevLett.73.752
26. C.F. Tejero, A. Daanoun, H.N.W. Lekkerkerker, M. Baus, Phys. Rev. E **51**, 558 (1995). https://doi.org/10.1103/PhysRevE.51.558
27. Z.T. Nemeth, C.N. Likos, J. Phys.: Condens. Matter **7**, L537 (1995), http://stacks.iop.org/0953-8984/7/i=41/a=002
28. C.N. Likos, G. Senatore, J. Phys.: Condens. Matter **7**, 6797 (1995), http://stacks.iop.org/0953-8984/7/i=34/a=005
29. C. Rascón, L. Mederos, G. Navascués, J. Chem. Phys. **103**, 9795 (1995). https://doi.org/10.1063/1.469944
30. G. Foffi, G.D. McCullagh, A. Lawlor, E. Zaccarelli, K.A. Dawson, F. Sciortino, P. Tartaglia, D. Pini, G. Stell, Phys. Rev. E **65**, 031407 (2002). https://doi.org/10.1103/PhysRevE.65.031407
31. P.G. Bolhuis, D. Frenkel, Phys. Rev. Lett. **72**, 2211 (1994), https://journals.aps.org/prl/abstract/10.1103/PhysRevLett.72.2211
32. P.G. Bolhuis, M. Hagen, D. Frenkel, Phys. Rev. E **50**, 4880 (1994), https://journals.aps.org/pre/abstract/10.1103/PhysRevE.50.4880
33. M. Dijkstra, R. van Roij, R. Evans, Phys. Rev. Lett. **81**, 2268 (1998). https://doi.org/10.1103/PhysRevLett.81.2268
34. K.N. Pham, A.M. Puertas, J. Bergenholtz, S.U. Egelhaaf, A. Moussaïd, P.N. Pusey, A.B. Schofield, M.E. Cates, M. Fuchs, W.C.K. Poon, Science **296**, 104 (2002). https://doi.org/10.1126/science.1068238

35. E. Zaccarelli, W.C.K. Poon, Proc. Natl. Acad. Sci. USA **106**, 15203 (2009), https://www.pnas.org/content/106/36/15203
36. C.P. Royall, S.R. Williams, H. Tanaka, J. Chem. Phys. **148**, 044501 (2018). https://doi.org/10.1063/1.5000263
37. C. Gögelein, R. Tuinier, Eur. Phys. J. E **27**, 171 (2008). https://doi.org/10.1140/epje/i2008-10367-6

Chapter 4
Unipletion in Colloid–Polymer Mixtures

4.1 Introduction

Mixtures of colloidal particles and polymers are widespread in biological [1–3] (e.g., blood) and industrial [4, 5] (e.g., paint) systems. Fundamental understanding of the colligative properties of colloid–polymer mixtures (CPMs) in terms of the molecular parameters involved provides a design pathway towards the desired final application. Particularly, the solvent quality [6, 7] as well as the specific polymer–colloid interaction [8] mediate the stability of the CPM. Polymers which do not adsorb onto the colloidal particles cause an indirect, *entropy-driven* attraction between the colloids, known as the depletion attraction [8]. On the other hand, polymer adsorption driven by *enthalpic* attractions between the polymer and the colloid leads to flocculation at low polymer concentrations and steric stabilisation at higher concentrations [9]. It is clear that both depletion and adsorption phenomena strongly mediate colloidal stability. Demixing is often undesirable since long-term stability is a requirement for products such as food and coatings. However, phase separation can be useful to fractionate compounds or extract certain components. Hence, continuous efforts take place for better understanding and controlling the stability of CPMs driven by polymer depletion or adsorption.

Most of the theoretical investigations on the phase behaviour of CPMs have been conducted on the depletion case, particularly under the assumption that the polymer concentration *at* the colloidal surface is strictly zero [10–12]. We term this situation *classical depletion*. Such a full depletion situation is formally restricted to a particular situation. Only if the effective interaction between the polymer segments and the surface is sufficiently repulsive there is a vanishing polymer segment concentration at the colloid [13], Whenever the entropic or enthalpic penalty for the presence of polymers near the colloidal surface is insufficient, there is a finite polymer concentration at the colloidal surface. Attention has been paid to understand the effect of this *weak depletion* as compared to the classical depletion case [6, 7, 14], also at high polymer concentrations [15]. Theory [16], experiments [6], and computer simulation

© Springer Nature Switzerland AG 2019

Á. González García, *Polymer-Mediated Phase Stability of Colloids*,
Springer Theses, https://doi.org/10.1007/978-3-030-33683-7_4

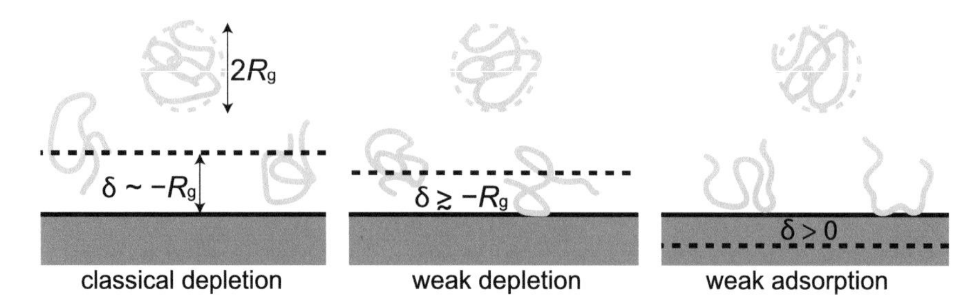

Fig. 4.1 Schematic representation of polymer configurations near a hard surface. Dashed lines represent the adsorption thickness δ, which is negative for depletion and positive for adsorption

[17] studies have revealed that the depletion thickness decreases at sufficiently high polymer concentrations.

Little attention has been paid to the colloidal interactions induced by *weakly adsorbing* polymers [12, 18]. For polymers which weakly (reversibly) adsorb onto the colloidal particle ('weak physisorption'), segment–surface attractions are of the order of the thermal energy. In this case, the phase behaviour of the CPM can be interpreted using thermodynamic descriptions such as the sticky hard sphere model [19, 20]. At low polymer concentrations, bridging attraction between the colloidal particles is expected because of the partial coverage of the colloids. In case the available amount of polymers is not sufficient to achieve (full) coverage of the colloidal surfaces, the polymer chains tend to adsorb onto two (or more) particles simultaneously. When the concentration of adsorbing polymer is high enough, the colloids become saturated with polymers, and hence restabilisation of the CPM may take place [21].

If the polymer affinity for the colloidal particle is sufficiently high, kinetic effects become relevant and polymers irreversibly adsorb at the colloidal particles [9, 22]. Both for depletion and bridging attraction, short-ranged and strong interactions may induce non-equilibrium phenomena. These include flocculation, aggregation, gelation [23], formation of percolated networks [18] and colloidal glasses [24]. These states are out of the scope of this chapter, where we present a theoretical framework for the *equilibrium* phase behaviour from classical depletion to weak adsorption (situations schematized in Fig. 4.1).

This chapter is inspired by classic studies of Scheutjens and Fleer on polymer-mediated interactions between two parallel plates [25]. Theoretically, we briefly revisit the well-established depletion and adsorption polymer concentration profiles near a hard surface for a dilute polymer solution. Then, we systematically vary the polymer–colloid affinity, which reveals a smooth transition from depletion to adsorption. In between we find a scenario where the polymer solution appears effectively unaffected by the colloidal particle. We term this condition unipletion, which reveals the possibility of preparing CPMs which are stable over a wide range of polymer concentrations. From the plate–plate interactions, we resolve the polymer-mediated pair potentials between colloidal hard spheres. Following the ideas of the extended law of corresponding states [26] allows us to construct the phase diagrams of CPMs ranging from classical depletion to weak adsorption *via* unipletion.

4.2 Interactions Between Hard Spheres: SCF and HCY

We follow the Scheutjens–Fleer self-consistent mean-field theory (SCF) for polymers at interfaces [21, 27, 28], and use the `sfbox` software (see Sect. 1.3.5) for obtaining homopolymer segment distributions and plate–plate interactions. As the SCF computations are based on Flory–Huggins mean-field theory, segment–segment interactions are captured via χ-parameters. We use a planar lattice with coordination number $k = 6$ (a simple cubic lattice), and consider concentration gradients in one dimension. The lower and upper boundary conditions of the lattice are user-defined. In this chapter, we consider two cases: (i) A mirror located after the last lattice layer combined with a surface impenetrable to all components in the system before the first lattice layer. (ii) An impenetrable surface both before the first and after the last lattice layer. In either case, the number of lattice sites is N_{lat}. In case (ii), N_{lat} corresponds to the distance (h) between the two hard plates. The [guest (G)] homopolymer chains added to the colloidal suspension are considered to be in a Θ-solvent (W). Hence, the polymer segment–solvent interaction is $\chi_{\mathrm{GW}} = 0.5$. The effective affinity of the polymer segments for the surface groups at a flat wall, which represent the surface of the colloidal (C) particle, is determined by the difference between polymer–colloid (χ_{CP}) and solvent–colloid (χ_{CW}) interactions. The interaction between the flat plate and a polymer segment (G) is set via $\Delta\chi = \chi_{\mathrm{CG}} - \chi_{\mathrm{CW}}$. At a fixed χ_{GW}, results are invariant at a fixed $\Delta\chi$ [6]. For simplicity, we set $\chi_{\mathrm{CW}} = 0$, so $\Delta\chi \equiv \chi_{\mathrm{CG}}$. All interactions are expressed in units of $k_B T$, with k_B the Boltzmann's constant and T the absolute temperature.

The SCF approach provides the polymer segment concentration profile that optimises the free energy of the lattice at the imposed $\phi_{\mathrm{G}}^{\mathrm{bulk}}$ and N_{lat}, following a semi-grand canonical approach at a fixed polymer bulk concentrations. The chemical potentials of all species in the bulk and in the system are equal. In the planar lattice, the resulting grand-canonical potential Ω comprises the free energy and the chemical potential of components in the lattice (see Sect. 1.3.5). Due to the Θ-solvent conditions, the (guest) homopolymer size may be characterised via its radius of gyration [29]:

$$R_{\mathrm{g}} = b\sqrt{\frac{N}{6}}, \tag{4.1}$$

with b the size of a lattice site and N the number of polymer segments. We set $N = 1000$ for all calculations in this chapter. As no long-ranged interactions (e.g., electrostatics) play a role and we normalize all distances either by the polymer size R_{g} or the colloidal diameter σ, there is no need to specify b. We express the homopolymer concentration ϕ_{G} relative to the overlap concentration ϕ_{G}^*, which satisfies:

$$\phi_{\mathrm{G}}^* = \frac{Nb^3}{v_{\mathrm{G}}} \approx 3.5N^{-1/2}, \tag{4.2}$$

with

$$v_{\rm G} = \frac{4\pi}{3} R_{\rm g}^3 \tag{4.3}$$

the volume occupied by a polymer coil in bulk at low concentrations. The interaction energy $W_{\rm plate}$ per unit area between two plates separated a distance h follows as [25]:

$$W_{\rm plate}(h) = \Omega(h) - \Omega(\infty). \tag{4.4}$$

We employ the Derjaguin approximation to obtain the interaction W between two spheres from the interaction between two plates [29]:

$$W(r) = W_{\rm HS}(r) + \frac{\pi\sigma}{2} \int_r^{\infty} W_{\rm plate}(r' - \sigma) {\rm d}r', \tag{4.5}$$

where r is the distance between the centres of the colloids and $W_{\rm HS}$ is the hard-sphere potential which accounts for impenetrability between the colloidal spheres (see Sect. 1.3.1). The size of the colloidal particle σ modulates the final pair potential beyond the hard-core. We impose a diameter σ such that $q = 0.15$, with

$$q \equiv \frac{2R_{\rm g}}{\sigma}. \tag{4.6}$$

The interaction potentials calculated via SCF (upon applying the Dejarguin approximation) are fitted to a HCY potential imposing the same second virial coefficient B_2 and area under the integral, considering spherical coordinates. Provided q, fitted ranges $q_{\rm Y}$ and strengths ϵ of the polymer-mediated effective colloid–colloid interaction are obtained as a function of $\phi_{\rm G}^{\rm bulk}$. This pair potential, the definition of B_2, and the free energy expressions applied to the fitted parameters were presented in Sect. 1.3.1–1.3.3.

4.3 Polymers Near Surfaces

We first consider polymers in a planar lattice where a single hard surface, which mimics the surface of the colloidal particle, is present. The resulting polymer concentration profiles already provide a first assessment of the polymer-mediated interactions between colloidal particles. For various values of the colloid–polymer affinity $\Delta\chi$, illustrative polymer segment concentration profiles near a single hard plate are plotted in Fig. 4.2 for a dilute polymer bulk concentration ($\phi_{\rm G}^{\rm bulk}/\phi_{\rm G}^* = 10^{-4}$). The expected depletion and adsorption homopolymer concentration profiles [21] near a hard wall are recovered for $\Delta\chi = +1.0$ and $\Delta\chi = -0.74$, respectively. For $\Delta\chi = +1.0$, classical depletion of the polymer from the colloidal surface takes place; $\phi_{\rm G}(z = 0) \approx 0$. For $z > 0$, the segment concentration increases up to its bulk

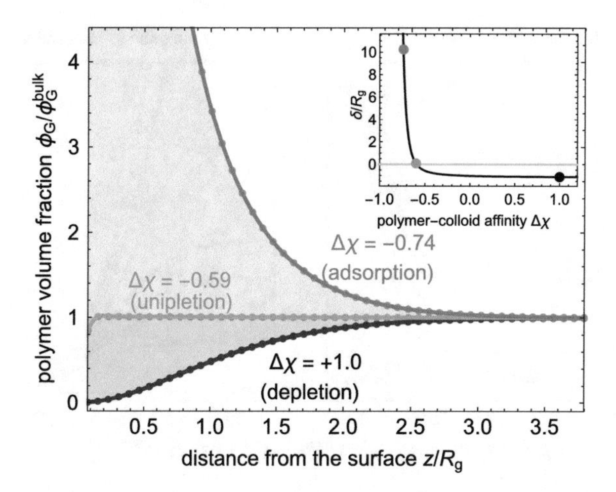

Fig. 4.2 Polymer concentration profiles relative to bulk (ϕ_G/ϕ_G^{bulk}) as a function of the distance from a flat surface (z/R_g). Polymer–colloid affinities $\Delta\chi$ indicated. In the inset, the adsorption thickness δ as a function of $\Delta\chi$ is presented. The situation $\delta \approx 0$ corresponds to the unipletion condition. Polymer bulk concentration relative to overlap is $\phi_G^{bulk}/\phi_G^* = 10^{-4}$. The polymer consists of $N = 1000$ segments with segment–solvent interaction $\chi_{GW} = 0.5$

value, reached around $z \approx 3R_g$. A precise evaluation for $\phi_G(z = 0) = 0$ corresponds to $\Delta\chi = 6\ln(7/6) \approx +0.92$ [30].

For adsorbing polymers ($\Delta\chi = -0.74$), the concentration near the colloidal hard surface is much greater than in bulk, and reaches its bulk value also around $z \approx 3R_g$. The equilibrium properties of adsorbed polymers are well-established [31], and the features of weakly-adsorbing polymers are well-recovered in SCF [21], hence they are not discussed here in more detail.

Remarkably, for $\Delta\chi \approx -0.59$ the polymer segment concentration profile is practically equal to ϕ_G^{bulk} up to the lattice site adjacent to the hard surface. This specific colloid–polymer affinity defines the *unipletion* condition, characterised by polymers that are neither depleted from nor adsorbed at the hard surface (see Fig. 4.1). In the inset of Fig. 4.2 we show the adsorption thickness δ of the polymer, which corresponds to the coloured areas of the main plot:

$$\frac{\delta}{b} = \sum_{z=1}^{N_{lat}} \frac{\phi_G(z)}{\phi_G^{bulk}} - 1. \tag{4.7}$$

Depletion is characterised by an adsorption thickness $\delta < 0$, while for adsorption $\delta > 0$. One could interpret δ as an effective amount of space gained ($\delta > 0$) or lost ($\delta < 0$) by the polymer due to the hard surface. The transition from depletion to adsorption occurs in a narrow $\Delta\chi$ range: the unipletion condition is characterised by vanishing δ. We denote this particular affinity as $\Delta\chi_u$. Note the non-linear behaviour of δ around $\Delta\chi_u$. Previously, it has been argued that homopolymer accumulation at the

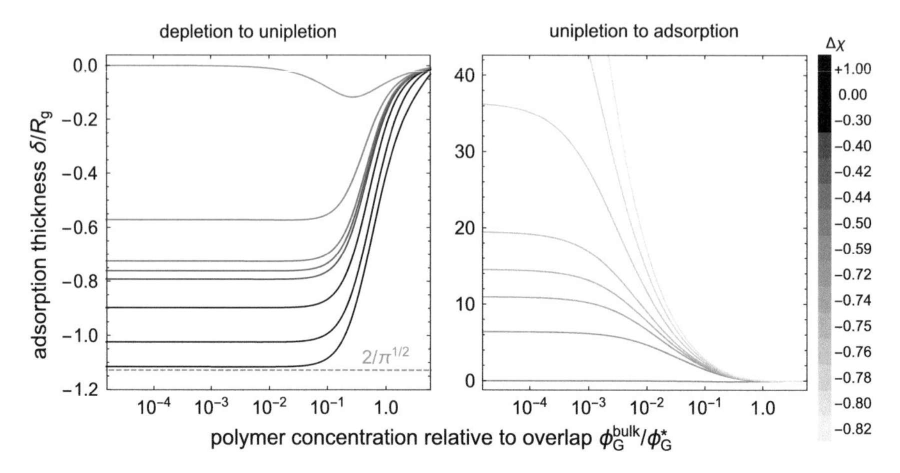

Fig. 4.3 Adsorption thickness δ as a function of the polymer bulk concentration relative to overlap ($\phi_{\mathrm{G}}^{\mathrm{bulk}}/\phi_{\mathrm{G}}^{*}$) for a collection of effective polymer–colloid affinities ($\Delta\chi$) as indicated. At $\Delta\chi \approx -0.59$, the unipletion condition holds, hence $\delta \approx 0$ up to relatively high polymer bulk concentrations

solid interface occurs at $\Delta\chi \lesssim \ln 5/6 \approx -0.18$ [32]. Contrary, we define depletion as the situation $\delta/R_{\mathrm{g}} < 0$ and adsorption for $\delta > 0$.

The polymer segment concentration profiles near a hard surface are modulated by the effective polymer–colloid affinity $\Delta\chi$, but also by the polymer bulk concentration $\phi_{\mathrm{G}}^{\mathrm{bulk}}$ and the polymer solvency χ_{GW} (the latter is not considered here). Ultimately, these density profiles also determine the polymer-mediated interactions between colloids. Independently of $\Delta\chi$, the osmotic pressure Π_{bulk} and consequently the fugacity of the polymers in bulk follows from Flory–Huggins theory [33]:

$$\widetilde{\Pi}_{\mathrm{bulk}} \equiv \frac{\Pi v_{\mathrm{G}}}{k_B T} = -\frac{N}{\phi_{\mathrm{G}}^{*}}\left[\ln\left(1-\phi_{\mathrm{G}}^{\mathrm{bulk}}\right) + \left(1-\frac{1}{N}\right)\phi_{\mathrm{G}}^{\mathrm{bulk}} + \chi_{\mathrm{GW}}\left(\phi_{\mathrm{G}}^{\mathrm{bulk}}\right)^2\right], \quad (4.8)$$

which for polymers in a Θ-solvent ($\chi_{\mathrm{GW}} = 0.5$) at low concentrations reduces to the Van 't Hoff law:

$$\widetilde{\Pi}_{\mathrm{bulk}} \approx \frac{\phi_{\mathrm{G}}^{\mathrm{bulk}}}{\phi_{\mathrm{G}}^{*}}. \quad (4.9)$$

In Fig. 4.3, the dependence of δ on $\phi_{\mathrm{G}}^{\mathrm{bulk}}$ is presented. For a polymer which is classically depleted from the surface we recover $\delta \approx -1.13R_{\mathrm{g}}$ as expected [13] for $\phi_{\mathrm{G}}^{\mathrm{bulk}}/\phi_{\mathrm{G}}^{*} \lesssim 0.05$ (see left panel). In all depletion cases ($1 \lesssim \Delta\chi \lesssim -0.59$, left panel of Fig. 4.3), δ is roughly independent of $\phi_{\mathrm{G}}^{\mathrm{bulk}}$ if $\phi_{\mathrm{G}}^{\mathrm{bulk}}/\phi_{\mathrm{G}}^{*} \lesssim 0.05$. For $\Delta\chi < 1$, the depletion thickness is always smaller than for the full depletion case due to a decreased free energy penalty of the guest polymer at the colloidal surface. For $\phi_{\mathrm{G}}^{\mathrm{bulk}}/\phi_{\mathrm{G}}^{*} \gtrsim 0.05$, the bulk osmotic pressure exerted by the polymers on the depletion zone (the volume where polymers are depleted) leads to compression [30, 34], and thus δ decreases in magnitude. This decrease of $|\delta|$ has been indirectly quantified experimentally for instance via AFM measurements [6].

In the case of homopolymer adsorption ($-0.59 \lesssim \Delta\chi \lesssim -0.82$, right panel of Fig. 4.3), the following trends are observed. A stronger effective segment–surface affinity (more negative $\Delta\chi$) results in a thicker adsorbed layer for all polymer bulk concentrations; i.e., a larger δ. As for the depletion attraction, δ eventually decreases in magnitude with increasing ϕ_G^{bulk}. Opposite to the depletion case, the range of constant δ depends on the specific $\Delta\chi$-value: the lower $\Delta\chi$, the lower the ϕ_G^{bulk} at which the surface is saturated with adsorbing polymers.

In case of unipletion ($\Delta\chi = \Delta\chi_u$), δ becomes negative around $\phi_G^{bulk}/\phi_G^* \approx 0.01$, see Fig. 4.3. A negative δ-value indicates the formation of a depletion layer. The origin of this decrease of δ lies in the saturation of the surface with unipleted polymer. Therefore the concentration of polymer at the surface is no longer proportional to the bulk concentration and δ decreases. Although a depletion layer is formed, this process effectively resembles the decrease of δ observed in adsorption. Further increase of the polymer concentration leads to compression of the depletion layers due to solvent removal, which resembles the increase of δ for depleting polymers. This highlights the dual and intricate nature of the unipletion condition, having both characteristics of depletion and adsorption.

We focus next on the concentration profiles in polymer solutions confined between two hard surfaces (Fig. 4.4). If the distance between the surfaces is large enough, the segment concentration profiles near either surface are like those reported in Fig. 4.2. For the depletion case, small interplate distances lead to a decrease in the maximum polymer concentration between plates with respect to the bulk due to an entropic penalty for polymers in confinement. In case of adsorbing polymers, the concentration between plates becomes larger than in bulk. The entropic loss of polymers in confinement is more than compensated by an enthalpic preference of the polymers towards the wall. Hence, when a free guest polymer chain can reach

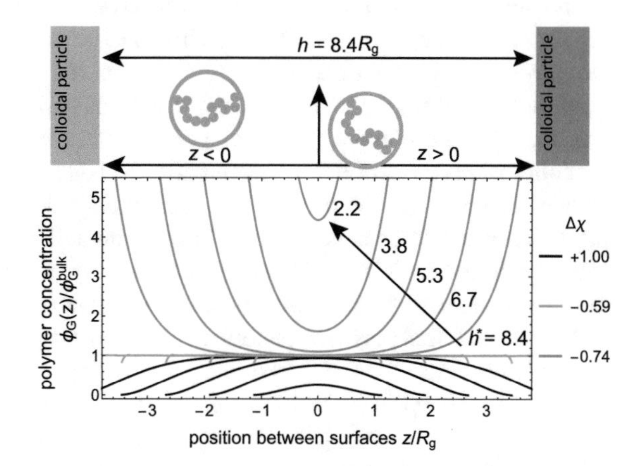

Fig. 4.4 Local polymer segment concentration profiles relative to bulk ϕ_G/ϕ_G^{bulk} as a function of the position z between two parallel, flat surfaces with decreasing the interplate-distance h (see sketch in the top panel). Effective polymer–colloid affinities $\Delta\chi$ are indicated. Arrow indicates decreasing inter-plate distance, with $h^* = h/R_g$

both surfaces ($h \approx 3R_g$, see Fig. 4.2) the polymer concentration is higher than in the bulk throughout the inter-plate region. Remarkably, in case of $\Delta\chi \approx -0.59$ the concentration profiles remain essentially flat for all interplate distances. The small dents observed at the hard surfaces relate to the fact that a polymer can not be positioned inside the hard wall.

4.4 Pair Interactions and Second Virial Coefficient

In this section, the interactions between colloidal spheres mediated by the addition of a guest homopolymer are studied at the two-body colloid–colloid interaction level. Examples of pair potentials between spherical colloids are shown in Fig. 4.5 at a low polymer bulk concentration ϕ_G^{bulk}. We focus first on classical depletion ($\Delta\chi = +1.0$), for which the attraction vanishes at $r/\sigma \approx 1 + q$. In this case, results are compared with the analytical expression by Eisenriegler [35]. The SCF data points, the fitted HCY curve, and the theoretical prediction are in agreement. The interaction beyond the hard-core for $\Delta\chi \approx -0.59$ is negligible: as polymers are neither adsorbed nor depleted at the colloidal surface, the colloidal particles interact effectively only via their excluded volume. When considering polymer adsorption ($\Delta\chi = -0.74$) at low polymer bulk concentrations, the resulting attraction is shorter in range but significantly stronger than the depletion attraction. Due to the preference of the polymers to sit at the colloidal surface, bridging-induced attraction occurs due to weakly adsorbing polymers at low concentration. The different ranges of the depletion and bridging attractions presented are magnified in the inset in the right panel of Fig. 4.5. For clarity, the fitted interactions are normalised by the absolute contact potential $|W(r = \sigma)|$. It follows that, for $\phi_G^{bulk}/\phi_G^* \lesssim 0.05$ the *normalised* depletion attraction remains fairly constant. Further details on the pair potentials with increasing ϕ_G^{bulk} and the results of the systematic HCY fit are presented in Sect. 4.7.

In a similar fashion as the adsorption thickness δ collects some key features of concentration profiles, we use the second virial coefficient B_2 to systematically quantify the pair–interactions. Furthermore, B_2 is often used as an indicator for the stability of a colloidal suspension [36–39]. If no interactions between spherical colloidal particles take place beyond their excluded volume, $B_2^* \equiv B_2/v_c = 4$, with $v_c = (\pi/6)\sigma^3$ the colloidal particle volume. Conveniently, the HCY model fit provides information on how the range and the strength of the interaction depend on ϕ_G^{bulk}. The discussion which follows is based not only upon the pair interactions, but also on the information extracted from the HCY fit. The collected ranges q_Y and strengths ϵ of the attraction from the systematic fit are also presented in Sect. 4.7.

We consider first polymer–colloid affinities ranging from depletion ($\Delta\chi = +1.0$) to unipletion ($\Delta\chi \approx -0.59$), see left panel in Fig. 4.6. Independently of the $\Delta\chi$-value, $B_2^* \approx 4$ if $\phi_G^{bulk}/\phi_G^* \lesssim 10^{-2}$. Further increase of ϕ_G^{bulk} first *decreases* B_2^* up to a minimum value around $\phi_G^{bulk}/\phi_G^* \approx 0.9$. The second virial coefficient then increases with ϕ_G^{bulk} back to $B_2^* \approx 4$. The depths of these minima in B_2^* decrease with decreas-

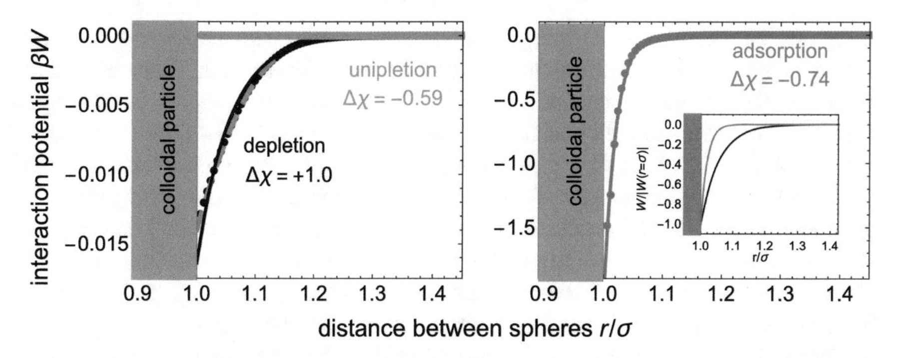

Fig. 4.5 Interaction between two spheres with diameter σ mediated by polymers with relative size $q \equiv 2R_g/\sigma = 0.15$ as a function of the distance between the centres of the spheres r. Effective polymer–colloid affinities $\Delta\chi$ are indicated. Solid curves correspond to the hard-core Yukawa (HCY) pair potential used to fit the calculated points. The dashed grey curve corresponds to theory by Eisenriegler [35] for the depletion attraction between two spheres. Polymer bulk concentration relative to overlap is $\phi_G^{bulk}/\phi_G^* = 10^{-3}$. Inset in the right panel shows the HCY-fitted depletion and adsorption potentials normalised by their absolute contact value $|W(r = \sigma)|$

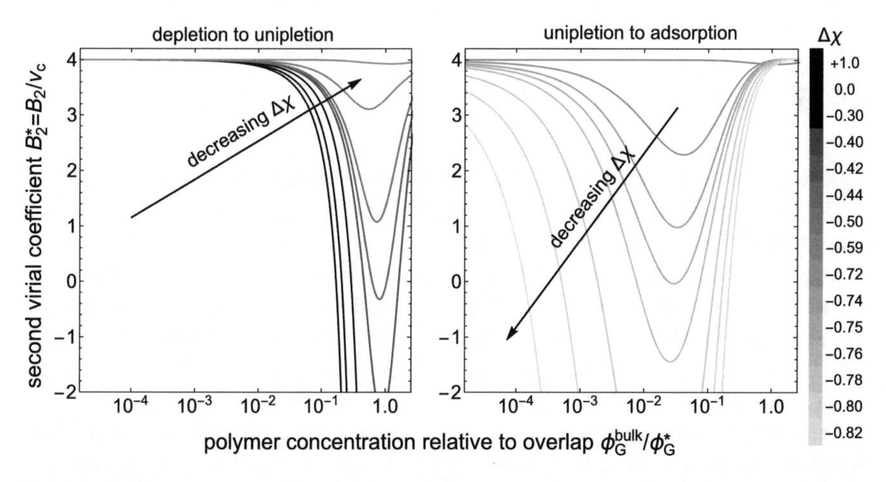

Fig. 4.6 Normalised second virial coefficient $B_2^* = B_2/v_c$ with increasing polymer bulk concentration ϕ_G^{bulk} relative to the polymer overlap concentration ϕ_G^*. The polymer–colloid affinities $\Delta\chi$ are indicated. For $\Delta\chi \approx -0.59$, unipletion is retained, and thus $B_2^* \approx 4$ for a wide range of polymer concentrations

ing $\Delta\chi$. The latter points towards a restabilisation of the colloidal suspension with decreasing $\Delta\chi$: the Vliegenthart–Lekkerkerker criterion [37] states that if B_2^* remains above -6, demixing of the CPM may be suppressed. Remarkably, $B_2^* \approx 4$ for all ϕ_G^{bulk} close to unipletion conditions ($\Delta\chi \approx -0.59$) even though δ is small yet finite at intermediate polymer concentrations ($0.01 \lesssim \phi_G^{bulk}/\phi_G^* \lesssim 1$, see Fig. 4.3, left panel).

In case of polymer adsorption onto the colloidal surface ($\Delta\chi \lesssim -0.59$), $B_2^* < 4$ already at very low polymer bulk concentrations. At a fixed, low ϕ_G^{bulk}, both the range and the strength of the bridging attraction *increase with decreasing* $\Delta\chi$ (see Sect. 4.7). For lower $\Delta\chi$, the polymers stretch from one surface to the other in order to maximize their overall contact with the colloidal surfaces. As expected, the stronger the polymer adsorbs to colloidal surface, the stronger the colloid–colloid attraction. For $\phi_G^{bulk}/\phi_G^* \lesssim 0.1$, both the range and the strength of the adsorption increase with ϕ_G^{bulk}. As for the depletion cases, B_2^* also reaches a minimum with increasing ϕ_G^{bulk}. In this case, as the colloidal surface fills up with polymer, competition between bridging attraction and steric (entropic) repulsion between adsorbed polymers takes place. A repulsive interaction beyond the hard-core occurs near and above ϕ_G^*, which results from a potential with a non-negligible attractive and very short-ranged strong repulsive contributions (see Fig. 4.8, bottom-right panel). Due to numerical limitations in these cases, we do not further discuss these effects here. The trends obtained for B_2^* point, also for weakly adsorbing polymers, towards a destabilisation–restabilisation transition around $\phi_G^{bulk} \approx 0.1$. Similar trends with increasing bridging agent concentration for the second virial coefficient have been recently reported for sticky hard sphere binary mixtures [19, 20]. Contrary to these approaches, we account specifically for the polymeric nature of the weakly adsorbing bridging agent.

4.5 Phase Diagrams

By virtue of the first order mean spherical approximation (FMSA), phase diagrams can be constructed from the fitted pair interactions obtained from the SCF approach combined with the Derjaguin approximation. Equal osmotic pressure $\widetilde{\Pi}$ and chemical potential $\widetilde{\mu}$ of the fluid and solid colloidal phases holds whenever colloidal coexistence takes place (see Chap. 1). Detailed results of the systematic fitting of the pair potentials can be found in Sect. 4.7. As the FMSA considers interactions added to the hard sphere (HS) reference state, the well-known fluid–solid coexistence for HSs [40] is recovered for all $\Delta\chi$-values in absence of polymer ($\phi_G^{bulk} = 0$).

Predicted phase diagrams are shown in Fig. 4.7. For polymers which are fully depleted from the colloidal hard surface [$\phi_G(z = 0) \approx 0$], the phase diagram matches predictions of the generalized free volume theory (GFVT) [29] for hard-spheres dispersed in a polymer solution at Θ-solvent conditions (see left panel of Fig. 4.7). Results from GFVT compare well with experimental results [10, 16, 29]. A significant decrease of $\Delta\chi$ is required in order to observe a shift of the fluid–solid binodal towards higher polymer concentration. Note how similar the phase diagrams are for $\Delta\chi = 1.0$ and $\Delta\chi = 0.0$. This corroborates the result that δ/R_g reaches a constant value with increasing $\Delta\chi$ (see inset of Fig. 4.2 for $\Delta\chi \gtrsim 0$). Decreasing $\Delta\chi$ further dramatically affects the fluid–solid binodal and increases the miscibility gap. For $\Delta\chi \approx 0.59$ the colloidal fluid–solid phase transition is only found at the coexisting concentrations expected for a polymer-free system. It is clear the concept of unipletion reveals the possibility of realising CPMs which are stable at high densities.

At a fixed colloid concentration and $\Delta\chi$ (e.g., $\phi_c = 0.15$ and $\Delta\chi = -0.4$), depletion destabilisation–restabilisation with increasing ϕ_G^{bulk} is revealed: with increasing

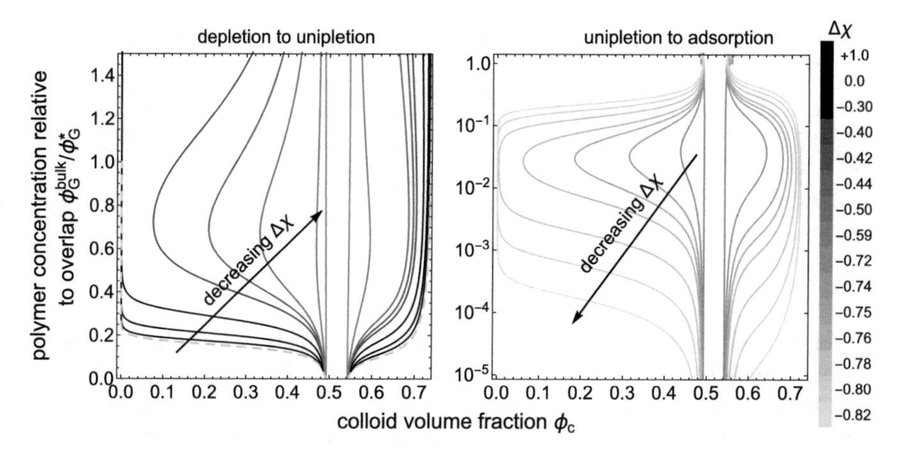

Fig. 4.7 Phase diagrams generated using the first order mean spherical approximation for the hard-core Yukawa fits of the SCF pair potentials. Dashed curve corresponds to the phase diagram obtained using generalized free volume theory under Θ-solvent conditions. For $\Delta\chi \approx -0.59$ unipletion takes place, and thus the colloidal fluid–solid coexistence remains at the same colloid volume fraction for all depletant concentrations

ϕ_G^{bulk}, the CPM goes from a stable one phase fluid to fluid–solid phase separation, and back to a single fluid phase (see Fig. 4.7, left panel). Similar trends have been observed in experiments where weaker depletion of polymer occurs due to the presence of short polymeric chains grafted to the colloidal particles [41, 42]. It has been argued that a slight repulsive bump in the colloidal pair interaction for a strongly-depleted polymer may be sufficient to explain this restabilisation [12, 43, 44]. Our SCF computations do reveal this repulsive bump in the depletion attraction at sufficiently high ϕ_G^{bulk}, whose magnitude is rather small compared to the contact potential (see Sect. 4.7). These tiny repulsive barriers [17] do not play a major role in our model for restabilisation.

Next, we pay attention to the phase diagrams obtained for weakly-adsorbing polymer (right panel of Fig. 4.7). Already at low ϕ_G^{bulk}, the (weakly) adsorbing polymers induce destabilisation of the colloidal suspension. As expected from the ϕ_G^{bulk}-dependence of both δ and B_2, weak adsorption-driven F–S demixing of the CPM occurs at higher polymer concentration with increasing $\Delta\chi$. Further, the lower $\Delta\chi$, the smaller the stable one-phase region. Still, even for relatively strong adsorption (e.g., $\Delta\chi = -0.8$) restabilisation occurs at high polymer concentration when the colloidal surfaces are saturated with adsorbing polymers. Contrary to the depletion case, with decreasing $\Delta\chi$ the adsorption thickness does not reach a limiting value. To the best of our knowledge, there are no previously-reported phase diagrams for *weakly adsorbing* polymers in CPMs where the nature of the bridging agent is taken into account. It is noted, however, that our computations are based upon thermodynamic equilibrium. In case of bridging effects the interactions are quite strong, and kinetic, non-equilibrium phenomena also become important [9, 22, 45].

Table 4.1 Depletion and weak-adsorption characteristics as captured with our computations at low polymer concentration

	$\Delta\chi$	$\phi_G(z=0)/\phi_G^{bulk}$	δ/R_g	$\beta W(r=\sigma)$	Range
Classic depletion	≈ 1	≈ 0	$=-2/\sqrt{\pi}$	$\propto \phi_G^{bulk}$	$=q$
Weak depletion	>-0.59	<1	$>-2/\sqrt{\pi}$	$\propto \phi_G^{bulk}$	$<q$
Unipletion	≈ -0.59	≈ 1	≈ 0	≈ 0	≈ 0
Weak adsorption	<-0.59	>1	>0	$\propto (\phi_G^{bulk})^n$	$\propto (\phi_G^{bulk})^n$

Yet another re-entrant phase behaviour can be extracted from our framework, namely upon varying $\Delta\chi$. Provided that there is an experimentally realizable tuning parameter for the effective polymer–colloid affinity, a transition from depletion to unipletion to adsorption is expected, which consequently changes the phase stability. Such a tuning parameter may for instance be the temperature; in fact the transition here described from depletion to adsorption by changing the temperature has been reported recently [46]. We must note that in real life also the solvency of the polymer changes with temperature, which may make the phase diagram transitions even richer than with the simple model presented here.

4.6 Conclusions

We present relatively simple computations which reveal the phase stability of colloid–polymer mixtures (CPMs) ranging from non-adsorbing to weakly-adsorbed polymers. For classical depletion conditions, the results from the well-established generalized free volume theory are recovered for the depletant-to-colloid size ratio $q = 0.15$. Near a specific effective polymer–colloid affinity (the unipletion condition), the CPM remains stable up to high polymer and colloid concentrations. The developed framework captures the different nature of the depletion and bridging attractions between colloidal particles, summarised in Table 4.1.

Furthermore, it follows both from the second virial coefficient and from phase diagrams that a destabilisation–restabilisation–destabilisation transition takes place as a function of the polymer bulk concentration when the colloid–polymer affinity is tuned from classical depletion to weak adsorption. The trends qualitatively match experimental observations on colloid–polymer mixtures.

4.7 Pair Potentials: Further Results

In Fig. 4.8, we present pair potentials between colloidal (hard) spheres for various polymer bulk concentrations. The HCY model accurately describes the SCF computations for various polymer concentrations. In the right panels we zoom on the range of small interaction energies. For the depletion case, the small repulsive bump

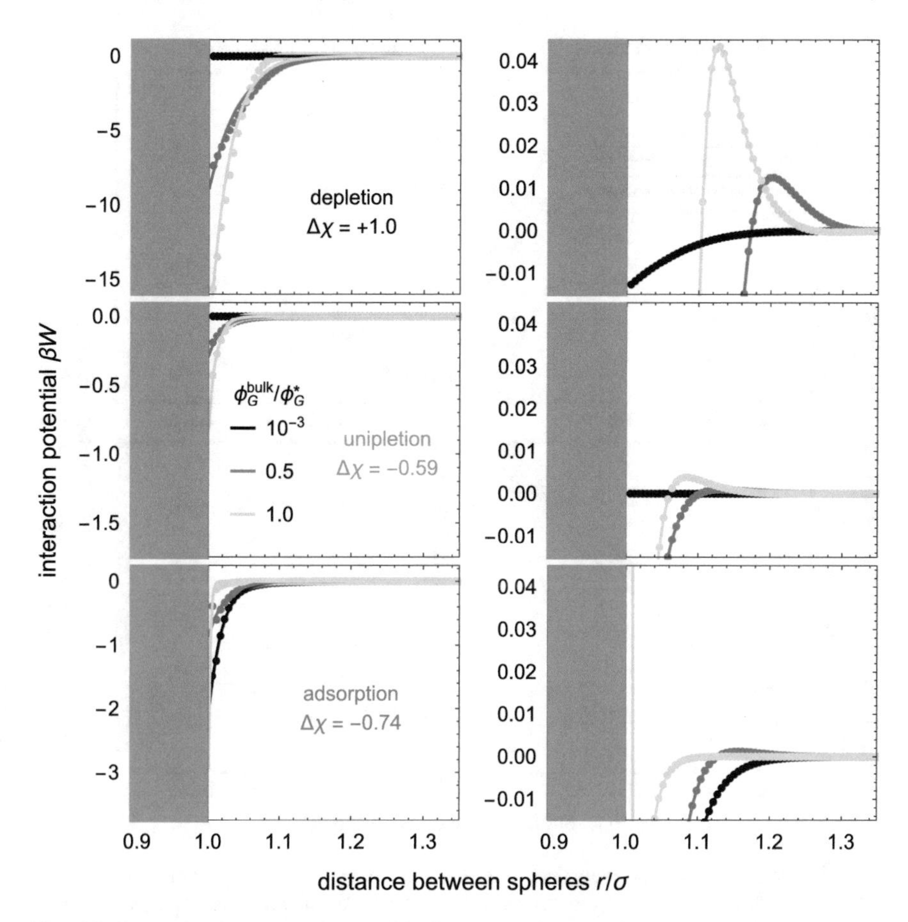

Fig. 4.8 Interaction between spheres with diameter σ due to polymers with relative size $q \equiv 2R_g/\sigma = 0.15$ as a function of the distance between the centres of the spheres r/σ for the polymer–colloid affinities $\Delta\chi$ indicated. In the left panels, solid curves correspond to the hard-core Yukawa (HCY) pair potential used to fit the calculated points. In the right panels, solid curves are to guide the eye. Polymer bulk concentrations relative to overlap are indicated. Right panels zoom on the range of small interaction energies for the potentials presented in the left panels

corresponds with the energy penalty of polymers escaping the depletion zone at high polymer concentrations. A tiny repulsive shoulder is also visible in case of unipletion at high polymer concentration. For the polymer adsorption cases, the strong and very short-ranged repulsion observed at $r/\sigma \approx 1$ may point towards steric repulsion between adsorbed polymers; however we do not further discuss these effects as our main focus is on dilute polymer solutions.

The results of the HCY-fits leading to the phase diagrams are presented in Fig. 4.9. The HCY-fits for the depletion cases show that q_Y is concentration-independent for the depletion attraction as expected in the dilute case ($\phi_G^{bulk}/\phi_G^* \lesssim 0.05$). Further, q_Y *decreases* above $\phi_G^{bulk}/\phi_G^* \approx 0.05$, corresponding to the compression of the depletion zones. On the other hand, the strength ϵ increases linearly with polymer concentra-

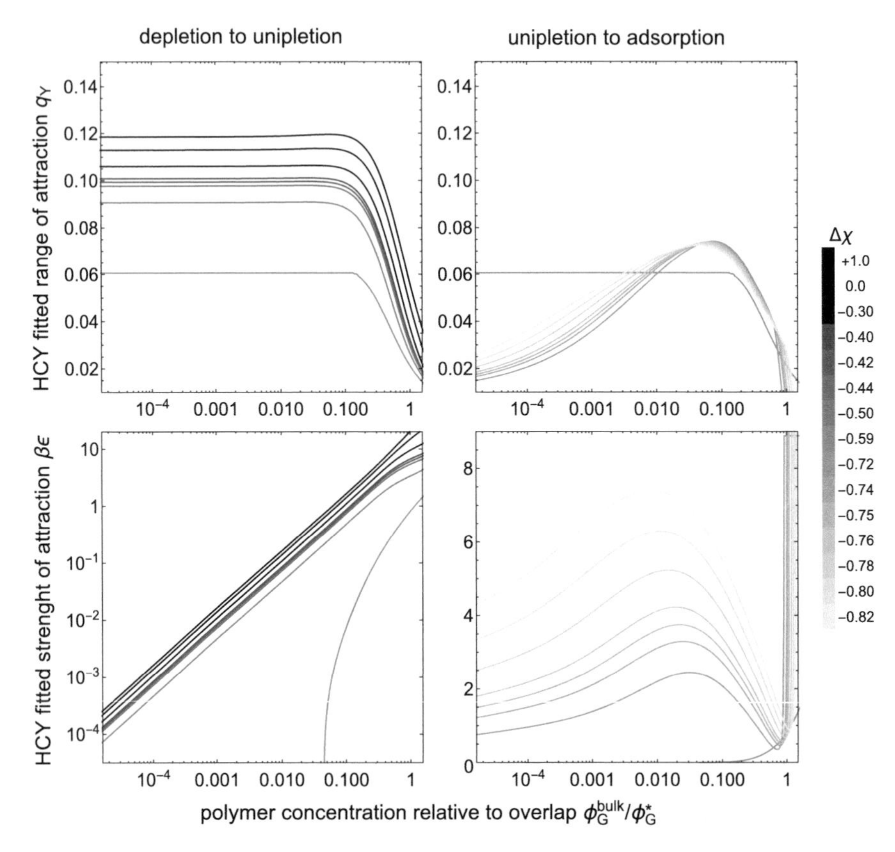

Fig. 4.9 Fitted ranges q_Y and strengths ϵ of the HCY model with increasing polymer bulk concentration ϕ_G^{bulk} relative to overlap ϕ_G^* for different effective colloid–polymer affinities $\Delta\chi$ as indicated

tion. Above $\phi_G^{bulk}/\phi_G^* \gtrsim 0.05$, the range decreases while the contact potential still increases with increasing ϕ_G^{bulk}.

In case of weak polymer adsorption, both the range q_Y and the strength ϵ of the bridging attraction increase with polymer bulk concentration for $\phi_G^{bulk}/\phi_G^* \lesssim 0.05$. Note that the maximum attraction strength shifts towards lower ϕ_G^{bulk} with decreasing $\Delta\chi$, again in accordance with what is observed for the adsorption thickness δ. The trends for adsorption around $\phi_G^{bulk}/\phi_G^* \approx 1$ point towards the limitations of fitting the SCF pair interactions with a single HCY potential. Phenomena such as the dramatic increase of ϵ accompanied by a decay of q_Y (observed in the right panels of Fig. 4.9) point towards potentials as the yellow curve in the bottom right panel of Fig. 4.8. From the trends on range and strength of interactions at low ϕ_G^{bulk}, the relevance of the unipletion condition becomes even clearer. While for depletion ϵ increases linearly and q_Y remains constant, for adsorption cases both ϵ and q_Y follow a power-law dependence with ϕ_G^{bulk}.

References

1. R. Mezzenga, P. Schurtenberger, A. Burbidge, M. Michel, Nat. Mater. **4**, 729 (2005), https://www.nature.com/articles/nmat1496
2. D. Marenduzzo, K. Finan, P.R. Cook, J. Cell. Biol. **175**, 681 (2006), http://jcb.rupress.org/content/175/5/681
3. E. Dickinson, Food Hydrocoll. **52**, 497 (2016), https://doi.org/10.1016/j.foodhyd.2015.07.029
4. T. Tadros, *Colloids in Paints* (Wiley, Hoboken,2011)
5. D. Saha, S. Bhattacharya, J. Food Sci. Technol. **47**, 587 (2010), https://doi.org/10.1007/s13197-010-0162-6
6. W.K. Wijting, W. Knoben, N.A.M. Besseling, F.A.M. Leermakers, M.A. Cohen Stuart, Phys. Chem. Chem. Phys. **6**, 4432 (2004), https://pubs.rsc.org/en/Content/ArticleLanding/2004/CP/b404030a#!divAbstract
7. S. Ouhajji, T. Nylander, L. Piculell, R. Tuinier, P. Linse, A.P. Philipse, Soft Matter **12**, 3963 (2016), https://pubs.rsc.org/en/Content/ArticleLanding/2016/SM/C5SM02892B#!divAbstract
8. X. Xing, L. Hua, T. Ngai, Curr. Opin. Colloid Interface Sci. **20**, 54 (2015), https://www.sciencedirect.com/science/article/pii/S1359029414001459
9. J. Gregory, S. Barany, Adv. Colloid Interface Sci. **169**, 1 (2011), https://www.sciencedirect.com/science/article/pii/S0001868611001229
10. D.G.A.L. Aarts, R. Tuinier, H.N.W. Lekkerkerker, J. Phys.: Condens. Matter **14**, 7551 (2002), http://iopscience.iop.org/article/10.1088/0953-8984/14/33/301/meta;jsessionid=98D0E74181CD42E623B408B6BCC0D2FC.c3.iopscience.cld.iop.org
11. M. Surve, V. Pryamitsyn, V. Ganesan, J. Chem. Phys. **122**, 154901 (2005), https://aip.scitation.org/doi/10.1063/1.1872772
12. A.N. Semenov, A.A. Shvets, Soft Matter **11**, 8863 (2015), https://pubs.rsc.org/en/content/articlelanding/2015/sm/c5sm01365h/unauth#!divAbstract
13. E. Eisenriegler, J. Chem. Phys. **79**, 1052 (1983), https://aip.scitation.org/doi/10.1063/1.445847
14. R. Tuinier, S. Ouhajji, P. Linse, Eur. Phys. J. E **39** (2016), https://link.springer.com/article/10.1140
15. J.B. Hooper, K.S. Schweizer, Macromolecules **39**, 5133 (2006), https://pubs.acs.org/doi/10.1021/ma060577m
16. G.J. Fleer, R. Tuinier, Adv. Colloid Interface Sci. **143**, 1 (2008), https://doi.org/10.1016/j.cis.2008.07.001
17. P.G. Bolhuis, A.A. Louis, J.P. Hansen, E.J. Meijer, J. Chem. Phys. **114**, 4296 (2001), https://doi.org/10.1063/1.1344606
18. M. Surve, V. Pryamitsyn, V. Ganesan, Langmuir **22**, 969 (2006), https://pubs.acs.org/doi/abs/10.1021/la052422y
19. J. Chen, S.R. Kline, Y. Liu, J. Chem. Phys. **142**, 084904 (2015), https://aip.scitation.org/doi/10.1063/1.4913197
20. R. Fantoni, A. Giacometti, A. Santos, J. Chem. Phys. **142**, 224905 (2015), https://aip.scitation.org/doi/full/10.1063/1.4922263
21. G.J. Fleer, M.A. Cohen Stuart, J.M.H.M. Scheutjens, T. Cosgrove, B. Vincent, *Polymers at Interfaces* (Springer, Netherlands, 1998), pp. XX, 496
22. B. O'Shaughnessy, D. Vavylonis, Eur. Phys. J. E **11**, 213 (2003), https://link.springer.com/article/10.1140/epje/i2003-10015-9
23. J. Bergenholtz, W.C.K. Poon, M. Fuchs, Langmuir **19**, 4493 (2003), https://pubs.acs.org/doi/abs/10.1021/la0340089
24. E. Zaccarelli, W.C.K. Poon, Proc. Natl. Acad. Sci. U.S.A **106**, 15203 (2009), https://www.pnas.org/content/106/36/15203
25. J.M.H.M. Scheutjens, G.J. Fleer, Adv. Colloid Interface Sci. **16**, 361 (1982), https://www.sciencedirect.com/science/article/pii/0001868682850252?via
26. M.G. Noro, D. Frenkel, J. Chem. Phys. **113**, 2941 (2000), https://aip.scitation.org/doi/10.1063/1.1288684

27. J.M.H.M. Scheutjens, G.J. Fleer, J. Chem. Phys. **83**, 1619 (1979), https://pubs.acs.org/doi/abs/10.1021/j100475a012
28. J.M.H.M. Scheutjens, G.J. Fleer, J. Chem. Phys. **84**, 178 (1980), https://pubs.acs.org/doi/abs/10.1021/j100439a011
29. H.N.W. Lekkerkerker, R. Tuinier, *Colloids and the Depletion Interaction* (Springer, Heidelberg, 2011)
30. G.J. Fleer, A.M. Skvortsov, R. Tuinier, Macromolecules **36**, 7857 (2003), https://pubs.acs.org/doi/abs/10.1021/ma0345145
31. V. Harabagiu, L. Sacarescu, A. Farcas, M. Pinteala, M. Butnaru, *Nanocoatings and Ultra-Thin Films*, eds. by A.S.H Makhlouf, I. Tiginyanu. Metals and Surface Engineering (Woodhead Publishing, Sawston, 2011) pp. 78–130, https://www.sciencedirect.com/science/article/pii/B978184569812650004X
32. A.A. Gorbunov, A.M. Skvortsov, J. van Male, G.J. Fleer, J. Chem. Phys. **114**, 5366 (2001), https://aip.scitation.org/doi/abs/10.1063/1.1346686
33. P.J. Flory, *Principles of Polymer Chemistry*, The George Fisher Baker Non-Resident Lectureship in Chemistry at Cornell University (Cornell University Press, Ithaca, 1953)
34. J. van der Gucht, N.A.M. Besseling, J. Phys.: Condens. Matter **15**, 6627 (2003), http://iopscience.iop.org/article/10.1088/0953-8984/15/40/002
35. E. Eisenriegler, Phys. Rev. E **55**, 3116 (1997), https://journals.aps.org/pre/abstract/10.1103/PhysRevE.55.3116
36. M.L. Kurnaz, J.V. Maher, Phys. Rev. E **55**, 572 (1997), https://journals.aps.org/pre/abstract/10.1103/PhysRevE.55.572
37. G.A. Vliegenthart, H.N.W. Lekkerkerker, J. Chem. Phys. **112**, 5364 (2000), https://aip.scitation.org/doi/10.1063/1.481106
38. A. Quigley, D. Williams, Eur. J. Pharm. Biopharm. **96**, 282 (2015), https://www.sciencedirect.com/science/article/pii/S0939641115003288
39. F. Platten, J.-P. Hansen, D. Wagner, S.U. Egelhaaf, J. Phys. Chem. Lett. **7**, 4008 (2016), https://pubs.acs.org/doi/abs/10.1021/acs.jpclett.6b01714
40. W.G. Hoover, F.H. Ree, J. Chem. Phys. **49**, 3609 (1968), https://aip.scitation.org/doi/10.1063/1.1670641
41. C. Cowell, R. Li-In-On, B. Vincent, J. Chem. Soc., Faraday Trans. 1 **74**, 337 (1978), https://pubs.rsc.org/en/Content/ArticleLanding/1978/F1/F19787400337#!divAbstract
42. B. Vincent, J. Edwards, S. Emmett, A. Jones, Colloids Surf. **18**, 261 (1986), https://www.sciencedirect.com/science/article/abs/pii/0166662286803171
43. A.N. Semenov, Macromolecules **41**, 2243 (2008), https://pubs.acs.org/doi/abs/10.1021/ma702536c
44. A.A. Shvets, A.N. Semenov, J. Chem. Phys. **139**, 054905 (2013), https://aip.scitation.org/doi/10.1063/1.4816469
45. T.G.V.D. Ven, Adv. Colloid Interface Sci. **48**, 121 (1994), http://www.sciencedirect.com/science/article/pii/0001868694800065
46. L. Feng, B. Laderman, S. Sacanna, P. Chaikin, Nat. Mater. **14**, 61 (2015), https://www.nature.com/articles/nmat4109#ref6

Part II
Anisotropic Hard Colloids

Chapter 5
Phase Behaviour of Colloidal Superballs Mixed with Non-adsorbing Polymers

5.1 Introduction

Colloidal cuboids are of interest due to their potential application as photonic crystals [1, 2], their possible roles in emulsion stabilisation [3] and anti-reflective coatings [4] and to prepare porous membranes [5]. Due to recent progress in colloidal synthesis, it is nowadays possible to prepare colloidal cuboids with a well-defined shape and size [6–8]. A commonly applied model to describe colloidal cuboids is the superball shape. Formally, superballs are a subset of a family of geometric shapes called superellipsoids, introduced by Barr [9]. The implicit equation describing the shape of a superball reads [10]:

$$\mathfrak{f}(x, y, z) = \left|\frac{x}{R}\right|^m + \left|\frac{y}{R}\right|^m + \left|\frac{z}{R}\right|^m \leq 1, \tag{5.1}$$

where R is the radius of the superball (the shortest distance from the centre of the superball to its surface, diameter $\sigma \equiv 2R$) and m is the shape parameter. The surface of the superball is described for $\mathfrak{f}(x, y, z) = 1$, whereas the locus of points inside the superball are retained for $\mathfrak{f}(x, y, z) < 1$. Here we focus on $m \geq 2$: the shape of the superball lies in between a sphere ($m = 2$) and a cube ($m = \infty$) [11]. We depict in Fig. 5.1 a collection of superballs in the range of m-values investigated.

The phase behaviour of colloidal superballs has been studied both experimentally [2, 12] and via computer simulations [11, 13, 14]. Some experimental studies on the effect of non-adsorbing polymers on the phase behaviour of colloidal superballs have also been conducted [6, 15]. However, closed-form equations for the thermodynamic properties of superballs (and superball–polymer mixtures) are not available. Hence, we first describe in this chapter a theoretical framework for both the superball fluid and for two possible superball solid states. We use free volume theory (FVT) to calculate the thermodynamic properties of superball–polymer mixtures. We present a collection of phase diagrams as well as a phase stability overview, which reveals various rich multi-phase coexistence regions, including a four–phase equilibrium.

© Springer Nature Switzerland AG 2019

Á. González García, *Polymer-Mediated Phase Stability of Colloids*,
Springer Theses, https://doi.org/10.1007/978-3-030-33683-7_5

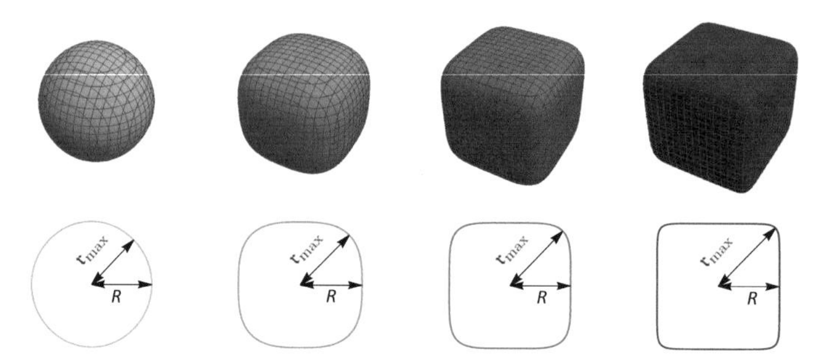

Fig. 5.1 3D (top panels) and 2D (bottom panels) representations of a superball for m-values (left to right): $m = \{2, 3, 5, 10\}$. In the 2D projection the radius of the superball (R) and the maximum distance of the superball surface from its centre ($\mathfrak{r}_{max}$) are indicated

5.2 Theory

In this section we first present the canonical expressions developed for a suspension of colloidal superballs, both for the fluid and the two solid phases considered. Secondly, we derive an expression for the excluded volume between a hard superball and a penetrable hard sphere (PHS), which renders the thermodynamic properties of a mixture of superballs plus non-adsorbing polymers.

5.2.1 *Canonical Thermodynamic Expressions for Colloidal Superballs*

Fluid State

We consider a collection of N_c hard superballs in a volume V, each superball having a volume v_c, surface area s_c and surface integrated mean curvature [16] c_c [see Eq. (5.28)]. The second virial coefficient (B_2) for hard particles is given by the orientationally-averaged excluded volume between two particles [17], and for a suspension of monodisperse convex particles (hence for superballs with $m \geq 2$) in a fluid state reads [18, 19]:

$$B_2/v_c \equiv B_2^* = 3\gamma + 1 \quad ; \quad \gamma = \frac{s_c c_c}{3v_c}, \tag{5.2}$$

where γ is the so-called asphericity parameter. The numerically obtained normalised second virial coefficient B_2^* for hard superballs is shown in Fig. 5.2. In Sect. 5.A we explain how γ can be computed numerically, yielding B_2^* using Eq. (5.2). It follows that B_2^* smoothly increases with m from the hard sphere (HS) limit ($m = 2$, $B_2^* = 4$) to the cube limit ($m = \infty$, $B_2^* = 5.5$) due to the increase of the particle anisotropy. In this work, we use a closed expression for B_2 by fitting (solid curve) the calculated

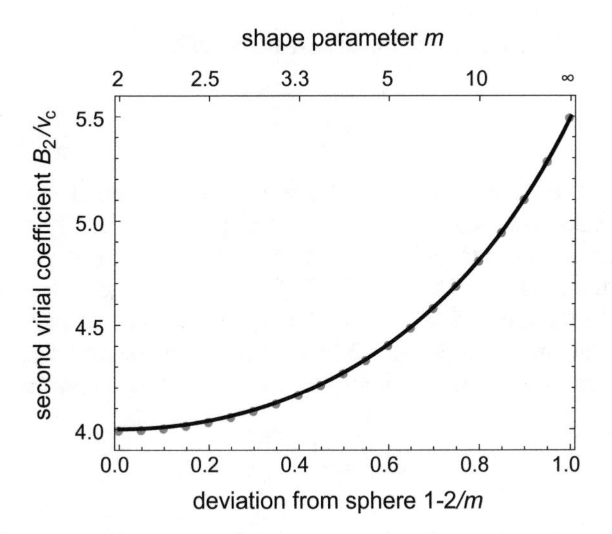

Fig. 5.2 Normalised second virial coefficient B_2^* as a function of the shape parameter m. Grey dots correspond to numerical solutions of the superball area and surface mean curvature (see Sect. 5.A for details). The black curve shows a fit through the data points, whose accumulated relative error is $5.15 \cdot 10^{-5} = 100\% * \sum [1/B_2^*(m) - 1/B_2^{*,\mathrm{fit}}(m)]/N_{\mathrm{points}}$, where N_{points} is the number of points used to find the fit in Eq. (5.3)

data with the (inverse) equation of an ellipse, which accurately matches the data (accumulated error shown in the caption of Fig. 5.2):

$$B_2^* \approx \frac{1}{0.42\sqrt{1-\left(\frac{1-2/m}{1.83}\right)^2}-0.17}. \tag{5.3}$$

An equation of state (EOS) for a fluid of hard convex particles was first derived by Gibbons using scaled particle theory [20], and a more accurate EOS was proposed by Boublík [21–23] taking into account virial coefficients higher than B_2 in a Carnahan–Starling-like fashion [24]:

$$\widetilde{\Pi}_{\mathrm{F}}^{\mathrm{o}} = \beta \Pi v_{\mathrm{c}} = \frac{\phi_{\mathrm{c}} + \mathfrak{Q}\phi_{\mathrm{c}}^2 + \mathfrak{R}\phi_{\mathrm{c}}^3 - \mathfrak{S}\phi_{\mathrm{c}}^4}{(1-\phi_{\mathrm{c}})^3}, \tag{5.4}$$

where $\widetilde{\Pi}_{\mathrm{F}}^{\mathrm{o}}$ is the reduced osmotic pressure of the pure hard superball dispersion, β is $1/k_B T$ (with k_B the Boltzmann constant and T the absolute temperature),

$$\mathfrak{Q} = 3\gamma - 2 \quad ; \quad \mathfrak{R} = 1 - 3\gamma(1-\gamma) \quad ; \quad \mathfrak{S} = \gamma(6\gamma - 5),$$

and where ϕ_c is the volume fraction of hard superballs:

$$\phi_c = \frac{N_c v_c}{V},$$

The superscript 'o' is used to indicate the (depletant-free) superballs system. Using the closed relation between B_2, γ and $\widetilde{\Pi}^o$ of Eqs. (5.2) and (5.4), the EOS for a fluid of hard superballs as a function of m is completely defined. Computer simulations have shown the accuracy of the Boublík EOS for a wide range of m-values [13, 14]. Obviously, the Carnahan–Starling [24] EOS is recovered for $m = 2$. The chemical potential of the superballs is related to the osmotic pressure through the Gibbs–Duhem relation for a single-component system at constant temperature:

$$\mathrm{d}\widetilde{\mu}^o_F = \frac{1}{\phi_c}\frac{\mathrm{d}\widetilde{\Pi}^o_F}{\mathrm{d}\phi_c}\mathrm{d}\phi_c, \tag{5.5}$$

with $\widetilde{\mu}^o = \beta\mu^o$ the reduced chemical potential. The chemical potential follows from Eqs. (5.4) to (5.5) as:

$$\begin{aligned}\widetilde{\mu}^o_F &= \widetilde{\mu}^{\mathrm{ref}} + \int_0^{\phi_c} \frac{1}{\phi_c}\frac{\mathrm{d}\widetilde{\Pi}}{\mathrm{d}\phi_c}\mathrm{d}\phi_c \\ &= \widetilde{\mu}^{\mathrm{ref}} + (\mathfrak{S} - 1)\ln(1-\phi_c) + \ln\phi_c \\ &+ \frac{(10 + 4\mathfrak{Q} + 2\mathfrak{S})\phi_c - (13 + 3\mathfrak{Q} - 3\mathfrak{R} + 5\mathfrak{S})\phi_c^2}{2(1-\phi_c)^3} + \frac{(5 + \mathfrak{Q} - \mathfrak{R} + \mathfrak{S})\phi_c^3}{2(1-\phi_c)^3},\end{aligned} \tag{5.6}$$

with $\widetilde{\mu}^{\mathrm{ref}} = \ln(\Lambda_B^3/v_c)$ the reference chemical potential of a superball fluid and Λ_B the thermal wavelength. The free energy follows from the chemical potential and the osmotic pressure through the thermodynamic relations:

$$\widetilde{F} = \phi_c\widetilde{\mu}^o - \widetilde{\Pi}^o \quad ; \quad \widetilde{\mu}^o = \left(\frac{\partial \widetilde{F}}{\partial \phi_c}\right)_{T,V}, \tag{5.7}$$

with $\widetilde{F} = \beta F v_c/V$ the reduced free energy.

Solid States

For the colloidal solid phases we modify the cell theory proposed by Lennard–Jones and Devonshire (LJD) for hard spheres (HSs) [25]. We consider each particle to be contained in a closed region whose shape is determined by its neighbouring particles, which are fixed at their lattice positions [26], see Fig. 5.3. The free energy of the solid is calculated from the number of configurations determined from the volume V_f that the centre of the particle explores without overlapping with its nearest neighbours. This leads to the following normalised free energy for a solid:

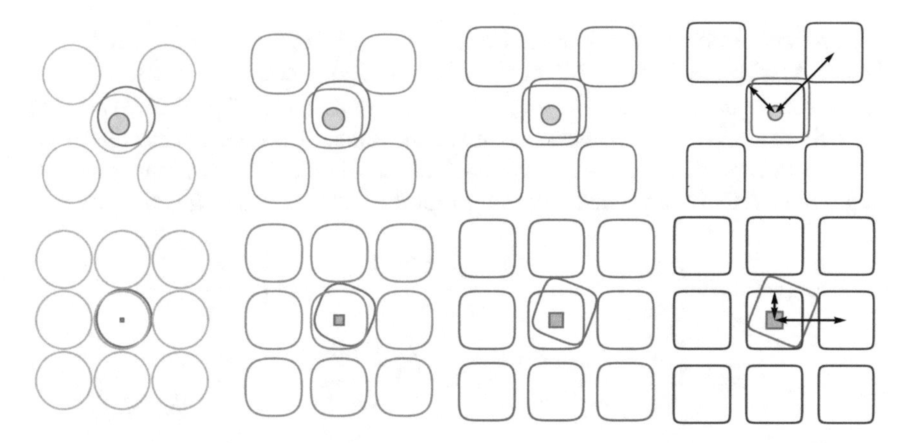

Fig. 5.3 Top panel: (100) plane representation of the face centred cubic (FCC) crystal lattice for increasing m-values from left to right: $m = \{2, 3, 5, 10\}$. The approximated free volume is illustrated as a grey region. Furthermore, a particle that just touches a nearest neighbour is also depicted. Bottom panels as top ones but for the simple cubic (SC) lattice. In all cases considered here the colloid volume fraction is $\phi_c = 0.45$. The arrows on the rightmost panels indicate the superball radius R and the nearest-neighbour distance r

$$\widetilde{F} = \phi_c \ln\left(\frac{\Lambda_B^3}{v_c}\right) - \phi_c \ln\left(\frac{V_f}{v_c}\right) . \tag{5.8}$$

The free volume V_f depends on the shape parameter m and the volume fraction ϕ_c, but also on the relative position of the nearest neighbours and hence on the structure of the solid. In this work we consider two crystal structures: face-centred cubic (FCC) and simple cubic (SC). A schematic view of the FCC and SC structures of superballs for several m-values is shown in Fig. 5.3.

We focus first on the FCC crystal. The exact free volume depends on the shape of the Wigner–Seitz cell [26], which for an FCC crystal has a rather complicated geometry [27, 28], but is usually approximated as a sphere [29]. The free volume considering this spherical approximation is given by:

$$V_f^{\text{FCC}} = \frac{4\pi}{3}\left(r - r_{\text{cp}}\right)^3 , \tag{5.9}$$

where r is the distance between the centres of a superball and its nearest neighbours, and r_{cp} is r at close packing. For the FCC crystal, we consider a 'frozen' crystal where the particles are perfectly aligned. Hence, for the FCC lattice r_{cp} is two times the distance between the edges of the superballs ($2\mathfrak{r}_{\text{max}}^{\text{2D}}$, see Sect. 5.A). The distance r at a certain volume fraction can be determined from r_{cp} as:

$$r = r_{\text{cp}}\left(\frac{\phi_c^{\text{cp}}}{\phi_c}\right)^{1/3} , \tag{5.10}$$

with ϕ_c^{cp} the close packing fraction. Combining $r_{cp} = 2\mathfrak{r}_{max}^{2D} = 2R\sqrt{2}\,(1/2)^{1/m}$ (see Sect. 5.A) with Eqs. (5.8)–(5.10) provides the free energy for the FCC phase. A Taylor expansion for the term $[(\phi_c^{cp,FCC}/\phi_c)^{1/3} - 1]$ is used as in the original LJD approach for HSs [25, 29]. The chemical potential and osmotic pressure are calculated via the thermodynamic relations given in Eq. (5.7), leading to the following closed (yet approximate) thermodynamic expressions for the FCC phase as a function of m:

$$
\begin{aligned}
\widetilde{F}_{FCC} &= \phi_c \ln\left(\frac{\Lambda_B^3}{v_c}\right) + \phi_c \ln\left[\frac{3^4 f(m)2^{3/m}}{4\pi 2^{3/2}}\right] - 3\phi_c \ln\left(\frac{\phi_c^{cp,FCC}}{\phi_c} - 1\right), \\
\widetilde{\mu}_{FCC}^{o} &= \widetilde{\mu}_0 + \ln\left[\frac{3^4 f(m)2^{3/m}}{4\pi 2^{3/2}}\right] - 3\ln\left(\frac{\phi_c^{cp,FCC}}{\phi_c} - 1\right) + \frac{3}{1 - \phi_c/\phi_c^{cp,FCC}}, \\
\widetilde{\Pi}_{FCC}^{o} &= \frac{3\phi_c}{1 - \phi_c/\phi_c^{cp,FCC}} .
\end{aligned}
\tag{5.11}
$$

The m-dependency of the close packing volume fraction in an FCC crystal is provided as (see Sect. 5.B):

$$
\phi_c^{cp,FCC} = \frac{1}{2} f(m) 2^{3/m}, \tag{5.12}
$$

where $f(m)$ is (see Sect. 5.A):

$$
f(m) = \frac{[\Gamma(1 + 1/m)]^3}{\Gamma(1 + 3/m)}, \tag{5.13}
$$

where Γ is the Euler Gamma function. For $m = 2$, Eq. (5.11) recovers the free energy for hard spheres in the FCC phase [25].

The thermodynamic properties of the SC superball crystal are found using a similar approach as for the FCC crystal. For the SC structure, the free volume has the shape of a cube. To approximately take into account the effect of rotations of cuboidal particles on the SC free energy, the size of the free volume is chosen such that the central particle can rotate inside its unit cell. This effectively reduces the free volume by a factor of eight compared to the one obtained for perfectly parallel cuboids. This approach overestimates the free volume for particles with low m-values. The free volume of the SC crystal is given by:

$$
V_f^{SC} = \left(r - r_{cp}\right)^3, \tag{5.14}
$$

with r defined via Eq. (5.10) and $r_{cp} = 2R$. Following a similar procedure as for the FCC phase state, the thermodynamic functions of the SC phase read:

Table 5.1 Close packing volume fractions for superballs with $m = 2$ (spheres), $m = 3$ and $m = \infty$ (cubes). At $m = 3$ both crystals have the same close packing fraction

	$m = 2$	$m = 3$	$m = \infty$
FCC	$\pi/3\sqrt{2} \approx 0.74$	$\Gamma(4/3)^3 \approx 0.71$	0.5
SC	$\pi/6 \approx 0.52$	$\Gamma(4/3)^3 \approx 0.71$	1

$$\widetilde{F}_{\mathrm{SC}} = \phi_{\mathrm{c}} \ln\left(\frac{\Lambda_{\mathrm{B}}^3}{v_{\mathrm{c}}}\right) + \phi_{\mathrm{c}} \ln f(m) - 3\phi_{\mathrm{c}} \ln\left[\left(\frac{\phi_{\mathrm{c}}^{\mathrm{cp,SC}}}{\phi_{\mathrm{c}}}\right)^{1/3} - 1\right],$$
$$\widetilde{\mu}_{\mathrm{SC}}^{\mathrm{o}} = \widetilde{\mu}_0 + \ln f(m) - 3\ln\left[\left(\frac{\phi_{\mathrm{c}}^{\mathrm{cp,SC}}}{\phi_{\mathrm{c}}}\right)^{1/3} - 1\right] + \frac{(\phi_{\mathrm{c}}^{\mathrm{cp,SC}}/\phi_{\mathrm{c}})^{1/3}}{(\phi_{\mathrm{c}}^{\mathrm{cp,SC}}/\phi_{\mathrm{c}})^{1/3} - 1}, \quad (5.15)$$
$$\widetilde{\Pi}_{\mathrm{SC}}^{\mathrm{o}} = \frac{\phi_{\mathrm{c}}(\phi_{\mathrm{c}}^{\mathrm{cp,SC}}/\phi_{\mathrm{c}})^{1/3}}{(\phi_{\mathrm{c}}^{\mathrm{cp,SC}}/\phi_{\mathrm{c}})^{1/3} - 1},$$

with the close packing fraction in the SC phase given by:

$$\phi_{\mathrm{c}}^{\mathrm{cp,SC}} = f(m)\,. \quad (5.16)$$

We note here that effects of particle rotations are only qualitatively accounted for via our estimation of V_{f}. For nonaxisymmetric hard particles, Onsager-like theories [30] for crystalline phases are non-trivial due to the lack of a single reference axes for the inter-particle orientations. Already for biaxial hard particles no analytical solutions are found [31]. In Table 5.1 we provide the close packing volume fractions for perfect spheres ($m = 2$) and perfect cubes ($m = \infty$) and for a limiting intermediate case. While FCC packings are more efficient for small m, SC arrangements can pack closer for large m. It follows from Eqs. (5.12) to (5.16) that both the FCC and the SC phases have the same $\phi_{\mathrm{c}}^{\mathrm{cp}}$ at $m = 3$. Further details on $\phi_{\mathrm{c}}^{\mathrm{cp}}$ are provided in Sect. 5.B.

Most of the limitations in our model are those inherent to the cell theory used in the calculation of the free energy of the crystalline phases. Cell theory is known to give accurate results for FCC and SC crystals of *spherical* colloids [27, 28], but extending cell theory to other crystal structures is not straightforward due to the complex geometries of the space explored by the centres of mass of the particles. Already for a body-centred-cubic crystal of HSs, cell theory does not match with simulations [28]. Furthermore, the cell theory approach followed does not account for defects, which affect the fluid-crystalline phase transition of cuboidal hard particles [32, 33]. The accuracy of cell theory for *anisotropic* particles is a matter of debate. Especially, it is challenging to accurately account for rotational contributions (in particular for nonaxisymmetric particles) into the solid-phase partition function. More advanced theoretical approaches to properly describe the complex solid phases of superballs are rather involved and would make the description at hand less tractable (and most likely not involving a set of closed expressions).

5.2.2 Free Volume Theory for Cuboid–Polymer Mixtures

As explained in Chap. 1, the required ingredients for calculating the thermodynamic properties of colloid–polymer mixtures (CPMs) using FVT are the equations of state of the depletant-free system and the colloidal particle-depletant excluded volume (v_{exc}). The general FVT expression for hard-colloids mixed with PHSs (with radius δ) holds in this case:

$$\widetilde{\Omega} = \widetilde{F}_{\text{c}} - \frac{v_{\text{c}}}{v_{\text{d}}} \widetilde{\Pi}_{\text{d}}^{\text{R}} (1 - \phi_{\text{c}}) \exp[-Q_{\text{s}}] \exp\left[-\frac{v_{\text{d}}}{v_{\text{c}}} \widetilde{\Pi}_{k}^{\text{o}} \right], \tag{5.17}$$

with $\widetilde{\Pi}_{\text{k}}^{\text{o}}$ the depletant-free osmotic pressure in a phase state k (F, FCC, or SC). Contrary to other FVT approaches presented through this thesis, simple expressions can not be obtained for the shape-dependent term Q_{s} [see Eq. (1.22)]. An apparently simple expression is available for the excluded volume between a convex body and a sphere [16]. The excluded volume between a hard superball and a hard sphere reads:

$$\widetilde{v}_{\text{exc}} = 1 + \frac{1}{f(m)} \left[\frac{1}{2} \widetilde{s}_{\text{sb}} q + \pi \widetilde{c}_{\text{sb}} q^2 + \frac{\pi}{6} q^3 \right] \tag{5.18}$$

where $\widetilde{s}_{\text{c}} = s_{\text{c}}/\sigma^2$ and $\widetilde{c}_{\text{c}} = c_{\text{c}}/\sigma$, and where

$$q = \frac{\delta}{R}. \tag{5.19}$$

It turns out that Eq. (5.18) can not be solved analytically (see Sect. 5.A for details on the calculation of $\widetilde{s}_{\text{sb}}$ and $\widetilde{c}_{\text{sb}}$). Due to the linear relation between δ and q ($\delta = qR$), $\widetilde{v}_{\text{exc}}(\lambda)$ is simply obtained by taking $q \rightarrow \lambda q$ in Eq. (5.18). Via interpolation of $\widetilde{s}_{\text{sb}}$ and $\widetilde{c}_{\text{sb}}$ it is possible to obtain a (non-closed) expression for Eq. (5.18), hence completely defining the grand potential Ω of the FVT used. Alternatively, we found that the depletion zone is accurately described by a tilted dampening sinus function:

$$\begin{aligned} \widetilde{v}_{\text{exc}} \approx & \frac{1}{f(m)} (454.34 + 216.36p + 308.59q) \\ & + \frac{1}{f(m)} \exp[-5 \cdot 10^3 p + 8.18] \sin(0.06p + 0.09q - 3.01), \end{aligned} \tag{5.20}$$

with $p = 1 - 2/m$. An advantage of Eq. (5.20) over Eq. (5.18) is that it is a closed-form expression. Subsequently, α is incorporated in our calculations considering the general form of the shape-dependent term Q_{s} given in Sect. 1.3.4. By inserting Eq. (5.20) into Eq. (5.17), the thermodynamics of hard superball–polymer mixtures can be expressed in terms of closed (fitted) functions. Furthermore, the deviation between Eqs. (5.18) and (5.20) is very small:

$$100\% * \sum [1/B_2^*(m) - 1/B_2^{*,\mathrm{fit}}(m)]/N_{\mathrm{points}} = 2.7 \cdot 10^{-4},$$

with N_{points} being the number of points used to fit Eq. (5.20). Moreover, the final thermodynamic properties of superball–polymer mixtures as considered here do not depend on the approach followed for calculating $\widetilde{v}_{\mathrm{exc}}$.

5.3 Results and Discussion

In the present section, we first provide the results for an ensemble of hard superballs in absence of depletants. The free volume fractions for depletants in this system is then briefly discussed, which provides all components to understand the phase diagrams of the superball–polymer mixtures of interest. Based upon these phase diagrams, an overview of the multi-phase coexistences exhibited is presented in a single plot which compromises the stable isostructural coexistences present at each set of relative depletant size and superball shape.

5.3.1 Phase Diagram of Hard Superballs

The calculated phase diagram for a suspension of pure hard superballs is presented in Fig. 5.4 (left panel), and compared with more evolved simulation results (right panel). The well-known fluid-FCC coexistence for hard spheres [29, 34] (HSs) is recovered at $m = 2$, and it slightly shifts to higher densities with increasing m. As can be appreciated, this is in agreement with computer simulation results [11]. This shift can be attributed to a thermodynamically less favourable FCC state with increasing m. The forbidden region (in grey) simply identifies densities beyond the close packing of the considered phases. The discontinuity along the border of the forbidden region is located at $m = 3$, which corresponds to the transition between preferred FCC to more favourable SC. The thermodynamically preferred phase is roughly the one with the largest close packing fraction, see Table 5.1. However, at intermediate pressures, crystal structures with lower close packing fraction may also be stable. We find a triple F–FCC–SC coexistence at $m \approx 3.71$. Triple phase coexistence for the hard superball system has also been reported [11] via Monte Carlo simulation results (right panel of Fig. 5.4). Between $m = 3$ and $m \approx 3.71$ we find SC–FCC coexistence, which arises from the different ϕ_{c}-dependencies of the solid equations of state (for more details, see Sect. 5.B). Above $m \approx 3.71$, only F–SC coexistence is found, which shifts towards lower packing fractions with increasing m, also in qualitative agreement with simulations [11]. In the cube limit ($m = \infty$), we find $\phi_{\mathrm{c}}^{\mathrm{F}} \approx 0.36$ and $\phi_{\mathrm{c}}^{\mathrm{SC}} \approx 0.54$. Computer simulation studies by Agarwal and Escobedo [14] indicate that for perfect cubes phase coexistence between a fluid and a cubatic liquid crystal takes place at $\phi_{\mathrm{c}}^{\mathrm{F}} \approx 0.47$ (a phase showing high orientational

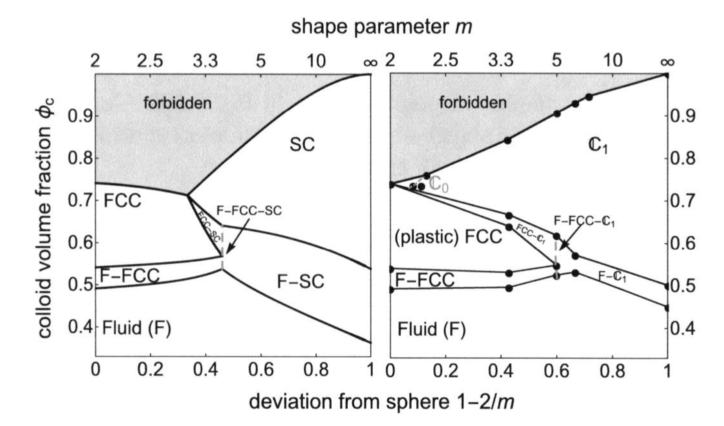

Fig. 5.4 Left panel: predicted phase diagram for a suspension of superballs in the shape parameter-colloid volume fraction $\{m, \phi_c\}$ phase space. Two-phase coexistences take place in the regions bounded by two single-phase regions as indicated. The vertical dashed grey line holds for the F–FCC–SC coexistence. Right panel: phase diagram for hard colloidal superballs from Monte Carlo computer simulations by Ni et al. [11]

order, but no long-range translational order), while at higher densities a transition into a SC crystal occurs at $\phi_c^{SC} \approx 0.58$. This points towards a complex nature of the F–SC phase transition for perfect cubes that can not be accounted for with our simple theory.

The overall topology of the theoretical phase diagram corresponds to the one found using more evolved computer simulations (right panel of Fig. 5.4) [11, 13, 14]. Differences can be justified because we do not account for the same solid phases for superballs as in simulations. Jiao et al. [10] showed that the packings of superballs are the $\mathbb{C}_0$ (low m-values) and $\mathbb{C}_1$ (high m-values) crystalline phases. Both the $\mathbb{C}_0$ and $\mathbb{C}_1$ lattices are obtained via deformation of the FCC ($m = 2$) and SC lattices ($m = \infty$), respectively. In fact, these solid phases are accounted for in simulations [11, 13]. Not surprisingly, the triple point from simulations is a fluid–plastic FCC–$\mathbb{C}_1$ [11]. Due to the limitations inherent to the simple theory used here, the $\mathbb{C}_0$ and $\mathbb{C}_1$ phases are not accounted for. The FCC phase features (and its coexistences) roughly match those of the plastic FCC. The role played in simulations by the $\mathbb{C}_1$ is mimicked by the SC phase in our simpler model.

When considering colloidal cuboid–polymer mixtures, the phase diagrams would only enrich upon refinements of the method. The liquid-crystalline and crystalline coexistence regions are found in simulations in a broader range of m-values. Based on experimental observations [6, 15], we expect the depletion attraction to enhance solid phases where depletion zones overlap is maximized (leaving space for the depletants to fit in the voids of the respective lattices). Note however that these experimental observations correspond to colloid–polymer mixtures confined at a surface, whereas the results here presented hold for bulk systems.

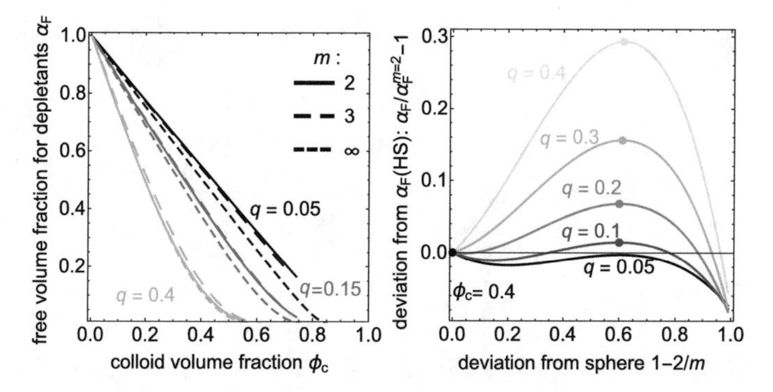

Fig. 5.5 Left panel: free volume fraction for PHSs in a colloidal superball fluid phase (α_F) as a function of the colloid volume fraction ϕ_c for the shape parameters m and relative depletant sizes q as indicated. For $q = 0.05$ the curves are plotted up to the corresponding highest close packing fraction. Right panel: relative difference of α_F with respect to spheres with increasing m at fixed $\phi_c = 0.4$ for a collection of q-values as indicated. Dots denote the maximum α_F-value at each m

5.3.2 Free Volume Fraction

We show examples of free volume fractions for PHSs in a fluid state of hard superballs (α_F) using $\widetilde{v}_{exc}$ as in Eq. (5.20) in the left panel of Fig. 5.5. It follows that α_F only weakly depends on m. For $m > 2$ and low q-values ($q \lesssim 0.05$) the free volume fraction is always slightly smaller than for the HS case ($m = 2$), see right panel of Fig. 5.5. However, with increasing q the intricate shape of the depletion zone around a superball causes α_F to be higher than for HSs. We do not pay further attention to the α in the solid phases considered. A geometrical free volume fraction for depletants in the different solid phases (Chap. 3) would improve the phase diagrams calculated, particularly for small q-values. However, already accounting for the most stable solid phases ($\mathbb{C}_0$ and $\mathbb{C}_1$) as a function of m would more significantly improve this first theoretical approach to superball–polymer mixtures.

5.3.3 Phase Diagrams

Firstly, we consider a superball whose shape is still close to a sphere. We present phase diagrams of superballs with $m = 2.5$ and added depletants for three relative size ratios q in Fig. 5.6. The depletant-free baselines ($\phi_d^R = 0$) for the fluid-FCC coexistence correspond to the densities shown in Fig. 5.4 (left panel). Upon addition of depletants, the FCC phase at coexistence gets denser and the coexisting fluid phase becomes more dilute in order to maximize the total free volume available for the depletants in the system. At sufficiently large q-values ($q = 0.4$ and $q = 0.6$ in Fig. 5.6), an isostructural colloidal F_1–F_2 (also termed gas–liquid) coexistence appears (which is metastable for low q-values). We do not further address metastable coexisting phases.

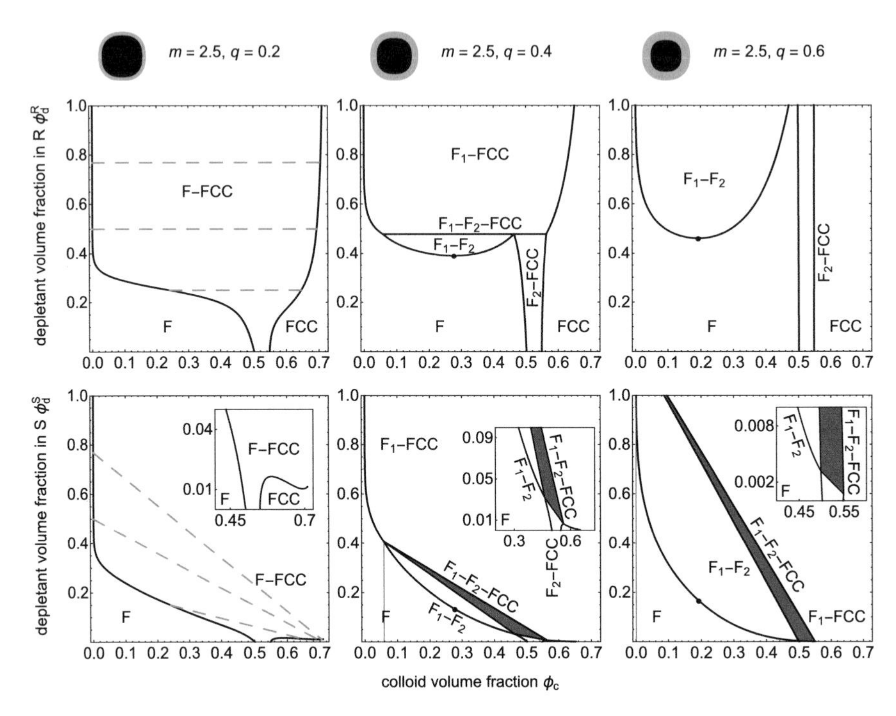

Fig. 5.6 Phase diagrams for a mixture of colloidal superballs for $m = 2.5$ and PHS depletants for several relative depletant sizes q as indicated in the reservoir representation (top panels) and system representation (bottom panels). The F_1–F_2 critical point is indicated by a black dot. Triple-phase coexistences are shown as a horizontal black line in the reservoir representation and as a coloured area bounded by black lines in the system representation. All coexisting phases present are indicated in the reservoir representation. Insets in the system representation zoom in on the low depletant concentration region, and some of the coexistence regions are indicated. A few illustrative tie-lines are shown as dashed grey lines for $q = 0.2$. Above each phase diagram, a 2D illustration of the superball (black) and its depletion zone (grey) are shown, and the m and q-values are indicated

This F_1–F_2 coexistence spans until the F_1–F_2 and the F–FCC coexistences match: at this ϕ_d^R-value a triple line is found (upper panel for $q = 0.4$), which becomes a region in the system representation (lower panels of Fig. 5.6). In fact, an isostructural phase coexistence is always connected to a triple coexistence when more than one phase identity is considered. When q increases, the F-S coexistence narrows and the F_1–F_2 critical point shifts to higher depletant volume fractions.

The coexistence regions in the system representation (bottom panels in Fig. 5.6) show that the fluid phase with a low concentration of superballs has a high concentration of depletants, whereas the FCC phase has a high concentration of superballs but a low concentration of depletants. The incorporation of partitioning of depletants over the different phases is one of the key elements of FVT [29]. The system representation also shows that for a superball-depletant mixture a single solid phase (*without* a coexisting fluid phase) only occurs at nearly imperceptible depletant concentrations.

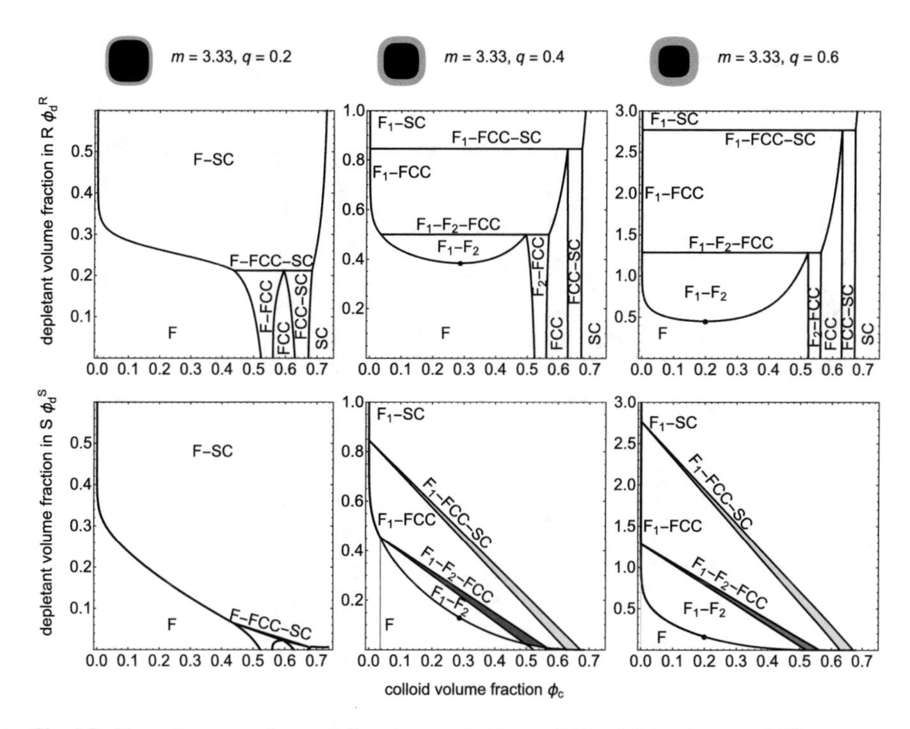

Fig. 5.7 Phase diagrams of superball–polymer mixtures as in Fig. 5.6, but for $m = 3.33$

So far we observe no special features compared to FVT for HSs mixed with PHSs [35], even though the colloidal shape considered is not perfectly spherical. For HSs plus polymeric depletants, simulation results taking into account multi-body effects show that the trends predicted by FVT hold [36].

Next, we show in Fig. 5.7 phase diagrams for $m = 3.33$, where FCC–SC coexistence was found for hard superballs (see Fig. 5.4, left panel). At sufficiently high depletant concentration, the FCC phase becomes metastable with respect to the SC phase, as expected because $\phi_c^{\text{cp,SC}} > \phi_c^{\text{cp,FCC}}$ for this m-value: the FCC phase completely disappears, and an F–FCC–SC triple point is always found. For sufficiently large q-values ($q = 0.4$ and $q = 0.6$) *two* triple-phase coexistences are present whenever F_1–F_2 coexistence is found: F_1–F_2–FCC and F–FCC–SC, with the corresponding triple point areas in the system representation.

Phase diagrams for even more cubic particles, $m = 5$, are presented in Fig. 5.8, for which only a SC state is present in the pure hard superballs system (see left panel of Fig. 5.4). Similar qualitative trends as in Fig. 5.6 are observed, but with the F–SC coexistence playing the role of the F–FCC equilibrium. For small q-values the broadening of the coexistence lines occurs at lower depletant concentrations with respect to the superball–polymer mixture with $m = 2.5$ in Fig. 5.6: the overlap of depletion zones is larger for particles with an increased cubicity, which results in a stronger depletion attraction. For $m = 5$ and $q = 0.4$, there is no F_1–F_2 equilib-

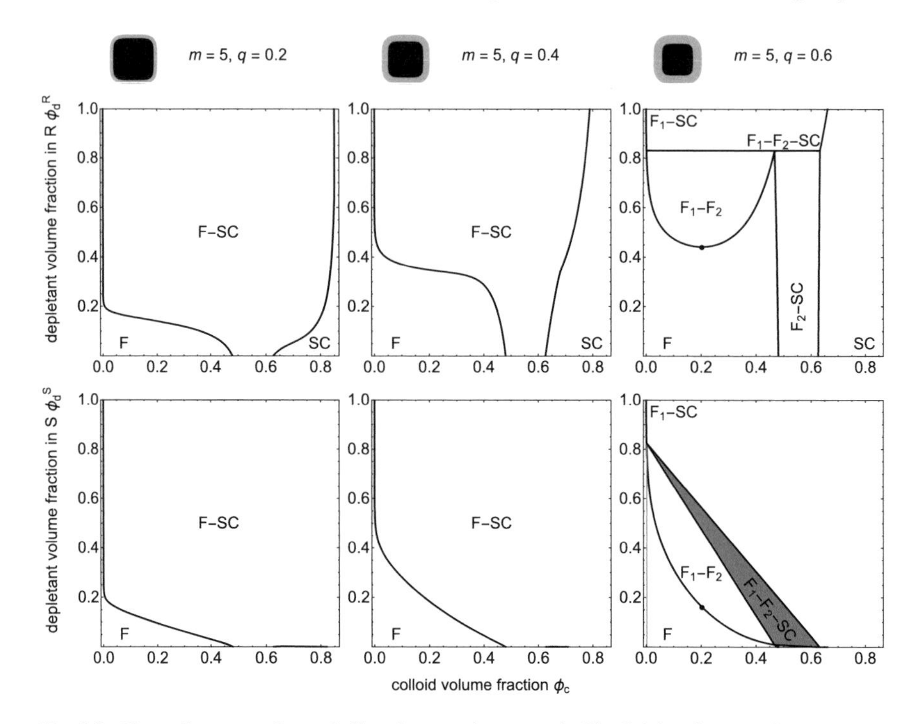

Fig. 5.8 Phase diagrams of superball–polymer mixtures as in Fig. 5.6, but for $m = 5$

rium phase coexistence, whereas F_1–F_2 coexistence was found at this q-value for superballs with $m = 2$ [35], $m = 2.5$ and $m = 3.33$. Due to the tendency of flat faces to align upon addition of depletant into the system, stable F_1–F_2 coexistence shifts to higher q-values: for larger m-values, longer ranges of attraction are required to induce a stable F_1–F_2 coexistence.

5.3.4 Multi-phase Coexistence Overview

The rich multi-phase coexistence behaviour hinted at in the previous section is quantified by calculating the critical end points (CEPs) of all possible isostructural coexistences as a function of the system parameters m and q. The calculated CEP curves are summarised in Fig. 5.9, which constitutes the main result of our investigations on the phase behaviour of superball–depletant mixtures. The limiting values of the FCC phase ($m \lesssim 3.71$) and the SC phase ($m \gtrsim 3$) are indicated as vertical dashed lines in Fig. 5.9: for $m \in \{3, 3.71\}$, F–FCC–SC coexistence always takes place. To the left of this m-interval the only solid phase found is FCC, and to the right the only solid state found is the SC. In Fig. 5.9, the solid curves hold for the CEPs defining the limiting q-values at which (stable) isostructural phase coexistences are found. For sufficiently

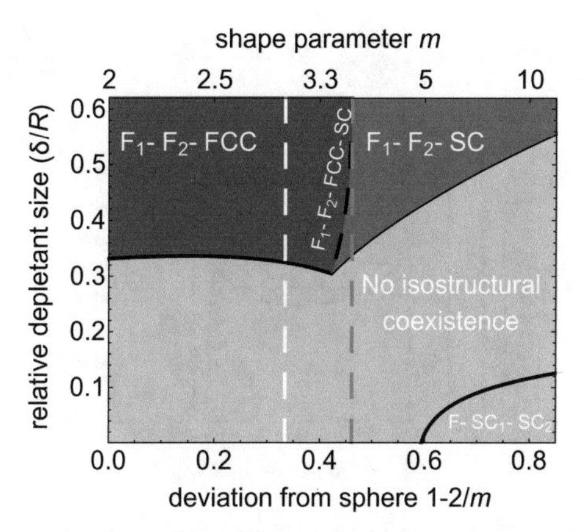

Fig. 5.9 Isostructural phase coexistence overview for colloidal superball–PHS mixtures as a function of the shape parameter m and the relative depletant size q. Isostructural coexistences correspond to coloured areas in the state diagram. The dashed black curve corresponds to the set of system parameters $\{m, q\}$ along which *quadruple* F_1–F_2–FCC–SC coexistence is found. Additionally, the vertical dashed white line corresponds to the limit of the SC phase ($m \geqslant 3$), whereas the grey vertical dashed line corresponds to the triple phase coexistence of the depletant-free system ($m \approx 3.71$), which sets the limit for the FCC phase ($m \lesssim 3.71$)

high q-values, F_1–F_2 isostructural coexistence is expected for all m-values, which spans from an F_1–F_2–FCC triple region or from an F_1–F_2–SC triple coexistence as indicated.

To gain insight about the F_1–F_2 isostructural coexistence and the connection with a triple F_1–F_2–FCC or F_1–F_2–SC coexistence we show in Fig. 5.10 phase diagrams for $q = 0.4$ and a few selected m-values (moving along a horizontal line in Fig. 5.9). For $m = 3.4$, a F–FCC–SC triple coexistence occurs at a higher $\phi_{\mathrm{d}}^{\mathrm{R}}$ than the F_1–F_2–FCC triple line (F_1–F_2–SC coexistence is metastable). On the other hand, for $m = 3.65$ the F_1–F_2–FCC triple point becomes metastable and an F–FCC–SC triple point arises at lower $\phi_{\mathrm{d}}^{\mathrm{R}}$ than the F_1–F_2–SC isostructural coexistence. This is explained by the fact that the stability of the FCC decreases as m increases. The condition at which the F_1–F_2–FCC and the F_1–F_2–SC coexistences merge results in a *quadruple* coexistence (F_1–F_2–FCC–SC). This four-phase coexistence is present for a range of m-values, and shows an asymptotic behaviour from the CEP of the quadruple coexistence towards the pure hard superball triple point (see Fig. 5.9). The shift of the QP from the TP of the depletant-free system towards lower m-values reflects the patchiness of the depletion attraction between superballs. As a consequence of the enhanced alignment of the flat faces upon addition of depletants, F–SC coexistence takes place at m-values below those of the depletant-free system. Hence, depletion-mediated entropic patchiness promotes the appearance of the SC phase. We conclude that quadruple coexistence arising from merging two isostructural triple phase coexistences is possible for superball–polymer mixtures. Such four-phase coexistences are possible in effective two-component systems provided there is an extra field variable:

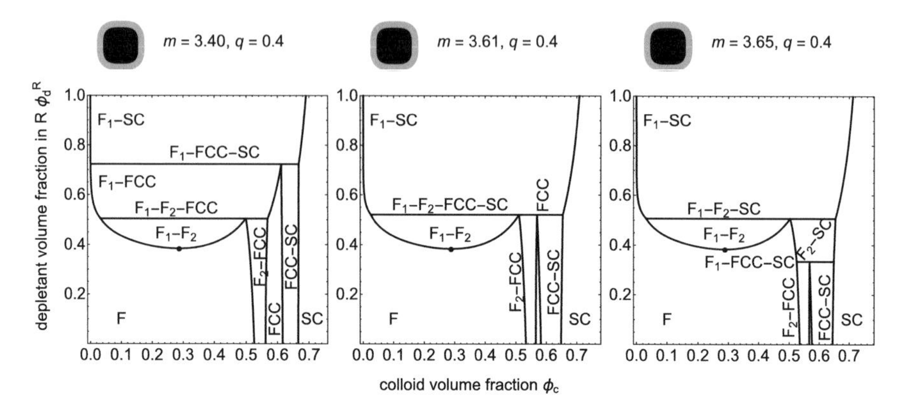

Fig. 5.10 Phase diagrams of mixtures of superball–polymer mixtures in the reservoir depletant concentration representation for $q = 0.4$ and various m-values as indicated

in this case, the superball's shape parameter m and the relative polymer size q act as these extra variables. As a F–FCC-$\mathbb{C}_1$ triple point has been detected in Monte Carlo computer simulation studies for the depletant-free superball system [11], the corresponding quadruple phase coexistence may be found from simulations with the $\mathbb{C}_1$ phase instead of the SC phase used here. However, hinting at this rich phase behaviour directly from simulations may be computationally demanding. The results provided by the simple model presented therefore are advantageous: they enable to map out possible phase coexistences efficiently.

A remarkable finding is the appearance of a SC_1–SC_2 isostructural coexistence found at low q and high m-values (see lower right part of Fig. 5.9). We depict a few illustrative phase diagrams in Fig. 5.11, where small isostructural SC_1–SC_2 coexistence regions appear. The single fluid phase and simple cubic regions get smaller upon decreasing q. For $m = 10$ the binodals shift towards lower ϕ_d^R-values with decreasing q, following the same trend as observed for the F_1–F_2, F–FCC and F–SC coexistences (see Figs. 5.6 and 5.8). This solid–solid coexistence is driven by the entropic gain for depletants upon phase separation of the colloids into a dense solid SC_2 phase and a more dilute SC_1 phase. The low q-values at which this coexistence takes place are related to the low (yet non-zero for cubic enough superballs) free volume fraction for depletants available in solid phases at high colloid concentrations. As can be observed in the rightmost panel of Fig. 5.11, the m-value tunes the depletant concentration at which SC_1–SC_2 coexistence is found: SC_1–SC_2 equilibria are driven by the alignment of the flat faces, and thus for more curved particles (decreasing m) the SC_1–SC_2 coexistence requires a higher depletant concentration. This induces the crystal state to demix into an attractive solid (depletion zones optimally overlapping) coexisting with a repulsive one, as expected for short-ranged attractions in colloidal systems [37, 38]. With more accurate models or in experimental systems, these SC_1–SC_2 coexistences may be replaced for example by a $\mathbb{C}_1$ coexisting with a SC phase. The absence of a stable FCC_1-FCC_2 coexistence can be rationalised by

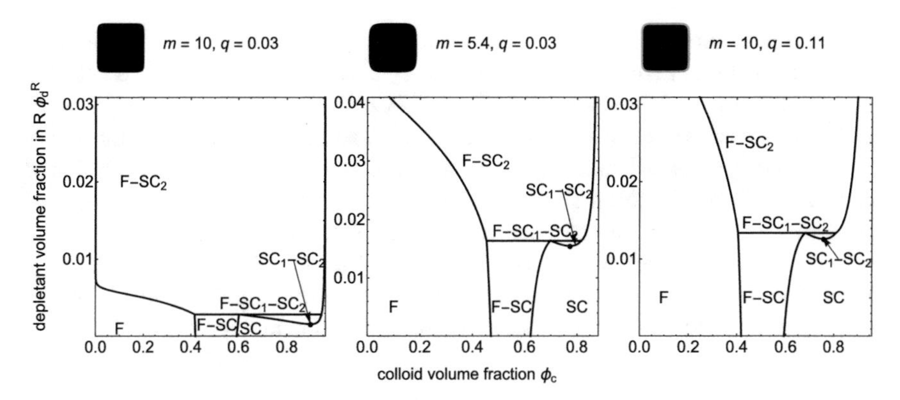

Fig. 5.11 Illustrative phase diagrams of superball–polymer mixtures in which isostructural SC phase coexistences appear

the non-optimal overlap of depletion zones between the flat faces of the superballs in an FCC state. However, in view of the results presented in Chap. 3, revision of the model proposed at small q-values is required to withdraw further conclusions.

5.4 Concluding Remarks

A simple model for the thermodynamic properties of superballs was presented, where all thermodynamic functions required for the phase diagram calculation are expressed in closed form. We account for three phase states: fluid (F), face centred cubic (FCC) and simple cubic (SC). Some of the closed expressions provided, such as the accurate fit for the second virial coefficient for hard superballs, may be of direct application not only in theoretical, but also in experimental studies. Despite the assumptions made, the found phase behaviour of pure hard superballs semi-quantitatively recovers the trends observed as compared with Monte Carlo (MC) computer simulations. Further improvements of the theory developed may be possible, particularly in the solid phases, but most likely lacking the simple, closed expressions reported here. The increase of the excluded volume with increasing particle anisotropy as well as the tendency for flat faces to align allowed us to rationalise the phase diagram obtained in terms of the colloidal packing fraction and the particle shape parameter.

The addition of free, non-adsorbing polymers to a collection of hard superballs induces effective attractive patches: alignment of the less-curved areas of the superballs is enhanced. This is clearly reflected in the presence of F–SC phase coexistence for shape parameters where it was not stable in the depletant-free system, and also in the manifestation of SC_1–SC_2 phase coexistence for sufficiently cubic colloids upon addition of small depletants. Such solid–solid coexistences may be not only of fundamental relevance, but also of relevance for designing novel photonic crystals. When the depletion attraction is sufficiently long-ranged, isostructural F_1–F_2 coexistence is found for all shape parameters, which can coexist either with an FCC or a SC state depending on the superball shape. The boundary between these two triple points that included isostructural fluid phases (F_1–F_2–FCC and F_1–F_2–SC) defines a window for *quadruple* phase coexistence (F_1–F_2–FCC–SC). The main trends were collected in a simple, comprehensive plot (Fig. 5.9) containing the system parameters (colloidal shape and relative depletant size) that summarises the effectively patchy nature of the depletion attraction in colloidal suspensions of cuboidal particles. The system parameters at which rich multi-phase behaviour of superballs–polymer mixtures is revealed with our simple model may guide more accurate computer simulations, and may serve as a first qualitative guide to interpret experimental results.

5.A Superball Properties: Calculation

Superball properties required for the calculation of the second virial coefficient and for the free energy of the solid phases are detailed in this section. The distance $\mathfrak{r}$ between the centre of a superball and an arbitrary point on the surface of the superball is given by [39]:

$$\mathfrak{r}(\theta, \phi) = R \left(|\cos\phi|^m |\sin\theta|^m + |\sin\phi|^m |\sin\theta|^m + |\cos\theta|^m\right)^{-1/m}, \tag{5.21}$$

where θ and ϕ are the polar angle and the azimuthal angle, respectively. The maximum distance ($\mathfrak{r}_{\text{max}}$) between the centre and the surface of the superball is the distance from the centre to the corner, as shown in Fig. 5.1 for the 2D superball projection. The angles corresponding to this maximum distance are $\theta = \pi/4$ and $\phi = 0$ for a 2D superball and $\theta = \arccos\left(\sqrt{2/3}\right)$ and $\phi = \pi/4$ for a 3D superball, which leads to a maximum distance given by:

$$\mathfrak{r}_{\text{max}}^{\text{2D}} = \mathfrak{r}\left(\frac{\pi}{4}, 0\right) = \sqrt{2}R\left(\frac{1}{2}\right)^{1/m}, \tag{5.22}$$

$$\mathfrak{r}_{\text{max}}^{\text{3D}} = \mathfrak{r}\left(\arcsin\left(\sqrt{2/3}\right), \frac{\pi}{4}\right) = \sqrt{3}R\left(\frac{1}{3}\right)^{1/m}. \tag{5.23}$$

The volume of the superball v_c is obtained by integration of Eq. (5.21) [39]:

$$v_c = \frac{8}{3}\int_0^{\pi/2}\int_0^{\pi/2} \sin(\theta) r(\theta, \phi)^3 d\theta d\phi, \tag{5.24}$$

where integration is performed over one octant due to symmetry. Equation (5.24) can be solved analytically, resulting in [5, 11]:

$$v_c = \sigma^3 f(m), \tag{5.25}$$

with

$$f(m) = \frac{[\Gamma(1+1/m)]^3}{\Gamma(1+3/m)}, \tag{5.26}$$

with σ the diameter of the superball ($\sigma = 2R$) and Γ the Euler Gamma function. Exact equations for the surface area s_{sb} and for the mean curvature c_{sb} of a superball are not known, but they can be calculated numerically using the surface integral and the integral of mean curvature [39]:

$$s_{\mathrm{c}} = 8 \int_0^{\pi/2} \int_0^{\pi/2} \mathrm{d}\theta \mathrm{d}\phi \left\| \frac{\partial \vec{x}}{\partial \theta} \times \frac{\partial \vec{x}}{\partial \phi} \right\|, \tag{5.27}$$

$$\begin{aligned} c_{\mathrm{c}} = & \frac{8}{4\pi} \int_0^{\pi/2} \int_0^{\pi/2} \mathrm{d}\theta \mathrm{d}\phi \\ & \Big\{ (\vec{x}_\theta \cdot \vec{x}_\theta)[(\vec{x}_\theta \times \vec{x}_\phi) \cdot \vec{x}_{\phi\phi}] + (\vec{x}_\phi \cdot \vec{x}_\phi)[(\vec{x}_\theta \times \vec{x}_\phi) \cdot \vec{x}_{\theta\theta}] - \\ & 2(\vec{x}_\theta \cdot \vec{x}_\phi)[(\vec{x}_\theta \times \vec{x}_\phi) \cdot \vec{x}_{\theta\phi}] \Big\} \times \\ & \Big\{ 2(\vec{x}_\theta \cdot \vec{x}_\theta)(\vec{x}_\phi \cdot \vec{x}_\phi) - 2(\vec{x}_\theta \cdot \vec{x}_\phi)^2 \Big\}^{-1}, \end{aligned} \tag{5.28}$$

where subscripts denote partial derivatives and $\vec{x}$ represents a vector from the centre of the superball to the surface, given by:

$$\vec{x} = \{\mathfrak{r}(\theta, \phi) \sin\theta \cos\phi, \mathfrak{r}(\theta, \phi) \sin\theta \sin\phi, \mathfrak{r}(\theta, \phi) \cos\theta\}, \tag{5.29}$$

with $\mathfrak{r}$ the distance from the centre of the superball (Eq. (5.21)). In view of the complicated forms of Eqs. (5.28) and (5.29), it is not surprising that formal solutions for the surface and mean curvature of superballs are not available.

5.B Close Packing and Free Volume of FCC and SC Crystals

In this section, clarification on the close packing fraction of the two solid states considered is provided. The general equation for the close packing fraction of a superball crystal is given by:

$$\phi_{\mathrm{c}}^{\mathrm{cp}} = \frac{N_{\mathrm{c}} v_{\mathrm{c}}}{V_{\mathrm{UC}}^{\mathrm{cp}}}, \tag{5.30}$$

with N_{c} the number of superballs in the crystal unit cell and $V_{\mathrm{UC}}^{\mathrm{cp}}$ the volume of the unit cell at the close packing fraction.

For the FCC crystal, the number of particles inside the unit cell is 4 and the volume of the unit cell at the close packing fraction is given by:

$$V_{\mathrm{UC}}^{\mathrm{FCC,cp}} = \left[4\mathfrak{r}_{\max}^{2\mathrm{D}} \sin\left(\frac{\pi}{4}\right)\right]^3 = (4R)^3 2^{-3/m},$$

which, combined with Eq. (5.30), gives the close packing fraction of superballs in the FCC crystal [Eq. (5.12)]. For the SC crystal, there is only a single particle inside the unit cell and the volume of the unit cell at the close packing fraction is simply given by:

$$V_{\mathrm{UC}}^{\mathrm{SC,cp}} = (2R)^3, \tag{5.31}$$

which results in the close packing fraction of superballs in a SC crystal given by Eq. (5.16).

References

1. Y.A. Vlasov, X.-Z. Bo, J.C. Sturm, D.J. Norris, Nature **414**, 289 (2001). https://doi.org/10.1038/35104529
2. J.-M. Meijer, A. Pal, S. Ouhajji, H.N.W. Lekkerkerker, A.P. Philipse, A.V. Petukhov, Nat. Commun. **8**, 14352 (2017). https://doi.org/10.1038/ncomms14352
3. J.W.J. de Folter, E.M. Hutter, S.I.R. Castillo, K.E. Klop, A.P. Philipse, W.K. Kegel, Langmuir **30**, 955 (2014). https://doi.org/10.1021/la402427q
4. B.G. Prevo, E.W. Hon, O.D. Velev, J. Mater. Chem. **17**, 791 (2007). https://doi.org/10.1039/B612734G
5. S.I.R. Castillo, D.M.E. Thies-Weesie, A.P. Philipse, Phys. Rev. E **91**, 022311 (2015). https://doi.org/10.1103/PhysRevE.91.022311
6. L. Rossi, S. Sacanna, W.T.M. Irvine, P.M. Chaikin, D.J. Pine, A.P. Philipse, Soft Matter **7**, 4139 (2011). https://doi.org/10.1039/C0SM01246G
7. J.R. Royer, G.L. Burton, D.L. Blair, S.D. Hudson, Soft Matter **11**, 5656 (2015)
8. F. Dekker, R. Tuinier, A.P. Philipse, Colloids Interfaces **2**, 44 (2018), https://www.mdpi.com/2504-5377/2/4/44
9. A.H. Barr, IEEE Comput. Graph. Appl. **1**, 11 (1981). https://doi.org/10.1109/MCG.1981.1673799
10. Y. Jiao, F.H. Stillinger, S. Torquato, Phys. Rev. E **79**, 041309 (2009). https://doi.org/10.1103/PhysRevE.79.041309
11. R. Ni, A.P. Gantapara, J. de Graaf, R. van Roij, M. Dijkstra, Soft Matter **8**, 8826 (2012). https://doi.org/10.1039/C2SM25813G
12. J.-M. Meijer, F. Hagemans, L. Rossi, D.V. Byelov, S.I. Castillo, A. Snigirev, I. Snigireva, A.P. Philipse, A.V. Petukhov, Langmuir **28**, 7631 (2012). https://doi.org/10.1021/la3007052
13. R.D. Batten, F.H. Stillinger, S. Torquato, Phys. Rev. E **81**, 061105 (2010). https://doi.org/10.1103/PhysRevE.81.061105
14. U. Agarwal F.A. Escobedo, Nat. Mater. **10**, 230 (2011). https://doi.org/10.1038/nmat2959
15. L. Rossi, V. Soni, D.J. Ashton, D.J. Pine, A.P. Philipse, P.M. Chaikin, M. Dijkstra, S. Sacanna, W.T.M. Irvine, Proc. Natl. Acad. Sci. U.S.A. **112**, 5286 (2015), https://www.pnas.org/content/112/17/5286?sid=c9bec40f-0b6f-4a2b-baca-636cfc3e3038
16. S.M. Oversteegen R. Roth, J. Chem. Phys. **122**, 214502 (2005). https://doi.org/10.1063/1.1908765
17. E. Herold, R. Hellmann, J. Wagner, J. Chem. Phys. **147**, 204102 (2017). https://doi.org/10.1063/1.5004687
18. A. Isihara T. Hayashida, J. Phys. Soc. Jpn. **6**, 40 (1951). https://doi.org/10.1143/JPSJ.6.40
19. H. Hadwiger, Experientia **7**, 395 (1951). https://doi.org/10.1007/BF02168922

20. R. Gibbons, Mol. Phys. **17**, 81 (1969), https://www.tandfonline.com/doi/abs/10.1080/00268976900100811
21. T. Boublík, Mol. Phys. **27**, 1415 (1974). https://doi.org/10.1080/00268977400101191
22. T. Boublík, J. Chem. Phys. **63**, 4084 (1975). https://doi.org/10.1063/1.431882
23. T. Boublík, Mol. Phys. **42**, 209 (1981), https://www.tandfonline.com/doi/abs/10.1080/00268978100100161
24. N.F. Carnahan K.E. Starling, J. Chem. Phys. **51**, 635 (1969), https://aip.scitation.org/doi/10.1063/1.1672048
25. J.E. Lennard-Jones, A.F. Devonshire, Proc. R. Soc A **163**, 53 (1937), https://www.jstor.org/stable/97067?seq=1#page_scan_tab_contents
26. M. Baus C.F. Tejero, *Equilibrium Statistical Physics*, 1st ed. (Springer, Heidelberg, 2008). https://doi.org/10.1007/978-3-540-74632-4
27. E. Velasco, L. Mederos, G. Navascués, Langmuir **14**, 5652 (1998). https://doi.org/10.1021/la980126y
28. S.K. Kwak, T. Park, Y.-J. Yoon, J.-M. Lee, Mol. Sim. **38**, 16 (2012). https://doi.org/10.1080/08927022.2011.597397
29. H.N.W. Lekkerkerker R. Tuinier, *Colloids and the Depletion Interaction* (Springer, Heidelberg, 2011)
30. L. Onsager, Ann. N. Y. Acad. Sci. **51**, 627 (1949). https://doi.org/10.1111/j.1749-6632.1949.tb27296.x
31. A. Cuetos, M. Dennison, A. Masters, A. Patti, Soft Matter **13**, 4720 (2017), https://pubs.rsc.org/en/Content/ArticleLanding/2017/SM/C7SM00726D#!divAbstract
32. F. Smallenburg, L. Filion, M. Marechal, M. Dijkstra, Proc. Natl. Acad. Sci. U.S.A. **109**, 17886 (2012), https://www.pnas.org/content/109/44/17886
33. A.P. Gantapara, J. de Graaf, R. van Roij, M. Dijkstra, Phys. Rev. Lett. **111**, 015501 (2013), https://journals.aps.org/prl/abstract/10.1103/PhysRevLett.111.015501
34. W.G. Hoover F.H. Ree, J. Chem. Phys. **49**, 3609 (1968), https://aip.scitation.org/doi/10.1063/1.1670641
35. H.N.W. Lekkerkerker, W.C.K. Poon, P.N. Pusey, A. Stroobants, P.B. Warren, Europhys. Lett. **20**, 559 (1992). https://doi.org/10.1209/0295-5075/20/6/015
36. M. Dijkstra, R. van Roij, R. Roth, A. Fortini, Phys. Rev. E **73**, 041404 (2006), https://journals.aps.org/pre/abstract/10.1103/PhysRevE.73.041404
37. P.G. Bolhuis, M. Hagen, D. Frenkel, Phys. Rev. E **50**, 4880 (1994), https://journals.aps.org/pre/abstract/10.1103/PhysRevE.50.4880
38. C.F. Tejero, A. Daanoun, H.N.W. Lekkerkerker, M. Baus, Phys. Rev. E **51**, 558 (1995). https://doi.org/10.1103/PhysRevE.51.558
39. D.J. Audus, A.M. Hassan, E.J. Garboczi, J.F. Douglas, Soft Matter **11**, 3360 (2015). https://doi.org/10.1039/C4SM02869D

Chapter 6
Discotic Dispersions Mediated by Depletion

6.1 Introduction

Colloidal suspensions and colloid–polymer mixtures (CPMs) in which the colloidal particles have a plate-like shape are common in nature and technology: examples are pigments [1], clays [2], blood [3], and foodstuffs [4]. Therefore the phase stability of colloidal platelets is of great relevance. Further, understanding their phase behaviour (also when mixed with other components) may enable separation of different components or enhance size fractionation. It has been experimentally observed how addition of non-adsorbing polymers [5–7] or small spheres [8–15] to a colloidal platelet suspension enriches the phase behaviour.

Theoretical models and computer simulations have predicted the experimentally observed isotropic–isotropic [5] (I_1–I_2) and nematic–nematic [16] (N_1–N_2) phase coexistences for platelets mixed with polymers or with small spheres. For infinitely thin plates, Bates and Frenkel [17] found I_1–I_2 and N_1–N_2 isostructural coexistences. Such I_1–I_2 and N_1–N_2 isostructural phase coexistences are a consequence of the depletion-induced attraction. Zhang et al. [18, 19] performed a computer simulation study and accounted for the finite platelet thickness, and incorporated the columnar phase. Alternative theoretical approaches regarding platelet–sphere mixtures have also reported such isostructural coexistences [20, 21]. Aliabadi et al. [22] presented a theoretical stability overview for hard plates in a sea of hard spheres (HSs).

In this chapter we show how the intricate effect of excluded volume interactions leads to a rich phase behaviour for platelet–polymer mixtures. The theoretical approach qualitatively recovers many results of previously-reported computer simulations and theories. On top of that, new coexisting phases are reported, namely a columnar–columnar (C_1–C_2) coexistence, and several types of three phase and four-phase coexistence regions, the latter being I_1–I_2–N–C, I–N_1–N_2–C, I–N–C_1–C_2. The structure of this Chapter is as follows. First, we report the basis of the theory developed, which provides the required tools to calculate the phase diagrams. Some results from our model are subsequently compared to those emerging from other more numerically involved methods and with simulation approaches. We summarise

© Springer Nature Switzerland AG 2019

Á. González García, *Polymer-Mediated Phase Stability of Colloids*,
Springer Theses, https://doi.org/10.1007/978-3-030-33683-7_6

our findings by presenting the various multiple phase equilibria that may occur in discotic colloid–polymer mixtures in a single, comprehensive plot spanned by the two relevant system parameters of our model, namely the disc aspect ratio and the colloid-to-polymer size ratio. The C_1–C_2 coexistence is studied in further detail, and results using a more precise account of the depletant partitioning in the columnar phase are compared with Monte Carlo (MC) simulations. Finally, we formulate the main conclusions.

6.2 Theory and Model Comparison

In this Section the Free Volume Theory (FVT) developed to study the colloidal platelet–polymer mixtures of interest is introduced. Firstly, we define the different length scales involved in our theoretical approach and recapitulate the known phase diagram for pure platelet suspensions. Subsequently, we show how a semi-grand canonical approach enables to compute the phase behaviour of platelet–polymer mixtures. For completeness, the equations of state of the pure platelet suspensions are presented in Sect. 6.A.

6.2.1 *Basics of the Model*

We model the colloidal platelets as discs with diameter σ and thickness L. The volume of the colloidal platelet (v_c) is given by $v_{\rm c} = \frac{\pi}{4}\sigma^2 L$. The aspect ratio $\Lambda = L/\sigma$ defines the shape of the colloidal particle, and we focus on $\Lambda < 0.3$. These colloidal platelets are described as hard particles: overlap leads to an infinite repulsive interaction while their interaction is zero otherwise. The polymeric depletants are simplified as penetrable hard spheres (PHSs) with radius δ (hence, their volume is $v_{\rm d} = \frac{4\pi}{3}\delta^3$, and δ corresponds to the depletion thickness [23]). The PHS concentration is expressed via the dimensionless concentration, $\phi_{\rm d}^{\rm R} = \rho_{\rm d}^{\rm R} v_{\rm d}$, with $\rho_{\rm d}^{\rm R}$ the number density of depletants in bulk. These PHSs can interpenetrate, but are hard for the colloidal discs. The PHS model is a good approximation for polymers at low concentration and in a θ–solvent [24]. The relative size of the depletant with respect to the colloidal platelet is defined in this Chapter as:

$$q \equiv \frac{2\delta}{\sigma}. \tag{6.1}$$

The three length scales introduced are schematically depicted in Fig. 6.1. The depletion zone surrounding any convex hard particle is enveloped by the surface with constant distance δ from the colloidal particle surface. Consequently, the volume of the depletion zone ($v_{\rm exc}$) corresponds to the excluded volume between a disc and a PHS [25]:

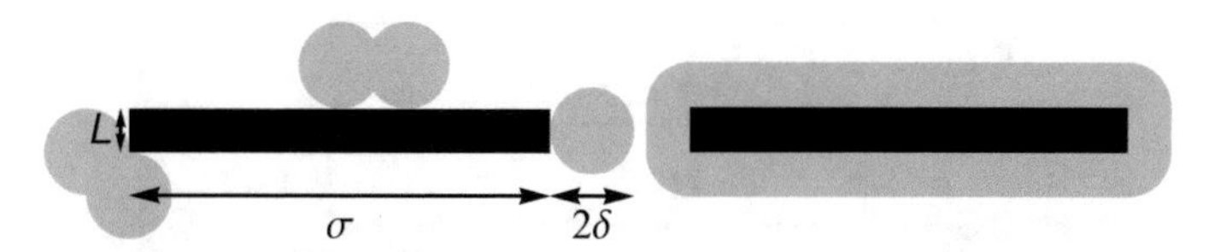

Fig. 6.1 Left: side view of a colloidal platelet with aspect ratio $\Lambda = L/\sigma$ and a few PHS depletants with relative size $q = 2\delta/\sigma$. The depletant diameter (2δ), the platelet diameter (σ), and the platelet thickness (L) are indicated. Right: corresponding depletion zone (grey) around the platelet (black)

$$v_{\text{exc}} = \frac{\pi}{2}\sigma^2\delta + \frac{\pi}{4}L(\sigma + 2\delta)^2 + \frac{\pi^2}{2}\delta^2\left(\sigma + \frac{8\delta}{3\pi}\right). \tag{6.2}$$

The typical shape of the depletion zone (side view) is depicted in Fig. 6.1.

We present our expressions for the free energy ($\widetilde{F}_k$, with subscript k running over the I, N and C states) of a system containing N_c hard discs in a volume V in terms of the volume fraction of platelets (ϕ_c), and use dimensionless units:

$$\phi_c = \frac{N_c v_c}{V} \quad ; \quad \widetilde{F}_k = \frac{\beta F_k v_c}{V}, \tag{6.3}$$

with $\beta = 1/k_B T$ in which $k_B T$ is the thermal energy with Boltzmann constant (k_B) and the absolute temperature T. Using the free energies for the different phase states, [26] standard thermodynamic relations can be applied to calculate the osmotic pressure (Π_k^{o}) and chemical potential (μ_k^{o}) of the pure platelet suspension in a given phase k (see Sect. 6.A).

This enables resolving the phase diagram for a system of hard platelets, presented in Fig. 6.2. The relatively high excluded volume between thin platelets explains the I–N phase transition occurring at very low packing fractions for very small values of the aspect ratio ($\Lambda \to 0$). With increasing Λ, the I–N coexistence widens and its boundaries shift towards higher packing fractions. From Fig. 6.2 it also follows that the N–C coexistence does not depend on Λ (within the model followed). For sufficiently thick discs ($\Lambda \approx 0.16$), transitions from an isotropic to a columnar phase occur without an intermediate nematic phase: thick discs are not sufficiently anisotropic to stabilize the occurrence of a nematic phase [26]. The grey vertical line in Fig. 6.2 at $\Lambda \approx 0.16$ indicates the I–N–C triple coexistence for hard colloidal platelets. The phase diagram presented in Fig. 6.2 constitutes the reference point for understanding the thermodynamics of platelet–polymer mixtures.

6.2.2 *Free Volume Theory for Platelet–Polymer Mixtures*

FVT is applied to compute the thermodynamic properties of colloidal platelets in a sea of polymer chains modelled as PHSs in a common solvent. Provided the depletant-free equations of state (Sect. 6.A), only the excluded volume between a plate and a

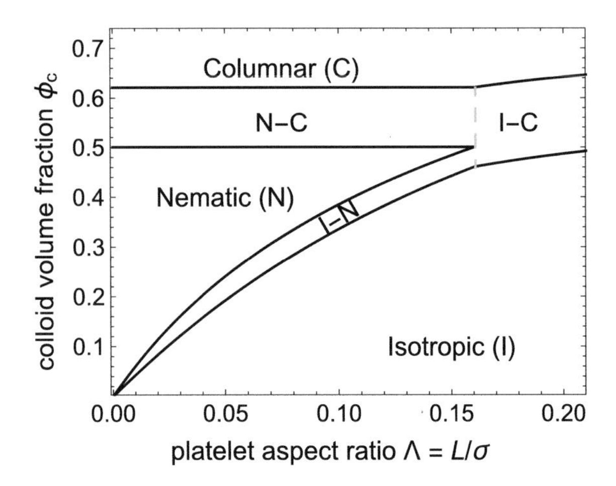

Fig. 6.2 Phase diagram of a monodisperse hard disc suspension in the aspect ratio–colloid volume fraction $\{\Lambda, \phi_c\}$ phase space. The dashed grey triple line indicates the Λ-value beyond which isotropic-columnar (I–C) coexistence dominates over the I–N and nematic-columnar (N–C), $\Lambda \approx 0.16$. At this specific Λ, there is an I–N–C triple point. A comparison of this phase diagram with computer simulation results from Bates and Frenkel [27] can be found in previous work by Wensink [26]. For more recent simulations, see Marechal et al. [28]

sphere is required to define the semi grand-potential Ω of the system:

$$\widetilde{\Omega} = \widetilde{F}_k - \frac{v_c}{v_d} \widetilde{\Pi}_d^R \alpha_k, \tag{6.4}$$

where

$$\alpha_k = (1 - \phi_c) \exp[-Q_s] \exp\left[-\frac{v_d}{v_c} \widetilde{\Pi}_k^o\right] \tag{6.5}$$

is the free volume fraction (calculated via scaled-particle theory) for depletants in a phase k, with

$$Q_s = q\left(\frac{1}{\Lambda} + \frac{\pi q}{2\Lambda} + q + 2\right) y + 2q^2 \left(\frac{1}{4\Lambda^2} + \frac{1}{\Lambda} + 1\right) y^2,$$

and

$$y = \frac{\phi_c}{1 - \phi_c}. \tag{6.6}$$

The depletant-free pressure of the phase state k considered is $\widetilde{\Pi}_k^o$. As we consider PHS depletants, the osmotic pressure of depletants in the reservoir R is given as:

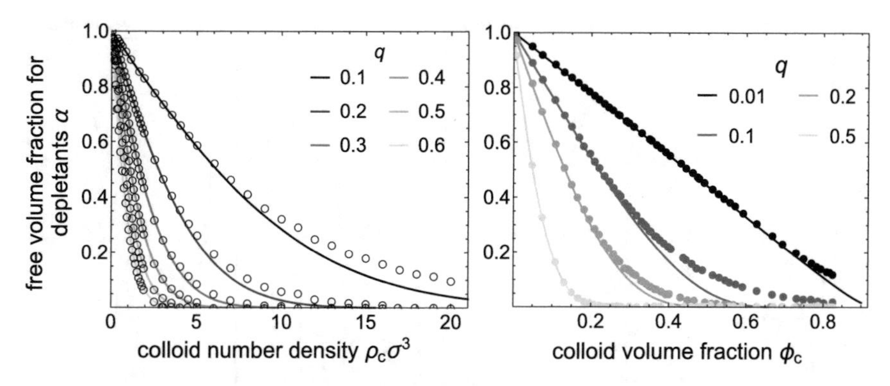

Fig. 6.3 Comparison of the free volume fraction using Eq. (6.5) and the osmotic pressure of an isotropic platelet phase [which follows from Eq. (6.14), curves] with available computer simulations literature data (symbols). Left: infinitely thin platelets (simulation data from Bates and Frenkel [17]) compared with $\Lambda = 10^{-5}$ in our approach. Right: cut-spheres with $\Lambda = 0.1$ (simulation data from Zhang et al. [19]). The relative size of the depletant q is indicated in the inset. Note that on the left panel we use the units as in the original reference

$$\frac{\Pi v_\mathrm{d}}{k_B T} \equiv \widetilde{\Pi}_\mathrm{d}^\mathrm{R} = \phi_\mathrm{d}^\mathrm{R}. \tag{6.7}$$

Details on the calculation of the phase diagrams are given in Sect. 1.3.4.

6.2.3 *Model Comparison*

We first compare our model both at the free volume level and at the phase diagram level with independent results. In Fig. 6.3 the calculated free volume fractions (using the osmotic pressure of the isotropic phase) from Eq. (6.4) are compared with previous results from computer simulations [17–19]. Our results are in concordance with simulation data both for infinitely thin plates [17] and for finite size cut-spheres [18, 19] mixed with PHSs. The agreement between our theoretical approach and previous simulations supports the validity of the derived expression for the free volume fraction of PHSs in a suspension of hard discs.

Phase diagrams for infinitely thin plates mixed with PHSs are compared (Fig. 6.4) with those determined by Bates and Frenkel using MC computer simulations [17]. We use dimensionless number density of platelets along the abscissa [$\rho_c\sigma^3 = 4\phi_c/(\pi\Lambda)$] and the fugacity of depletants ($\bar{z} = \rho_\mathrm{d}^\mathrm{R}\sigma^3 = \phi_\mathrm{d}^\mathrm{R} q^{-3} 6/\pi$) on the ordinate for comparison. In their approach, the free volume fraction for PHSs in a disc suspension is fitted from simulation results, and FVT is applied to the calculation of the phase diagrams considering the full numerical solution of the disc orientation probability function in the nematic phase. As the difference in the I–N coexistence between Odijk's Gaussian trial function approximation and the self-consistent numerical approach for the

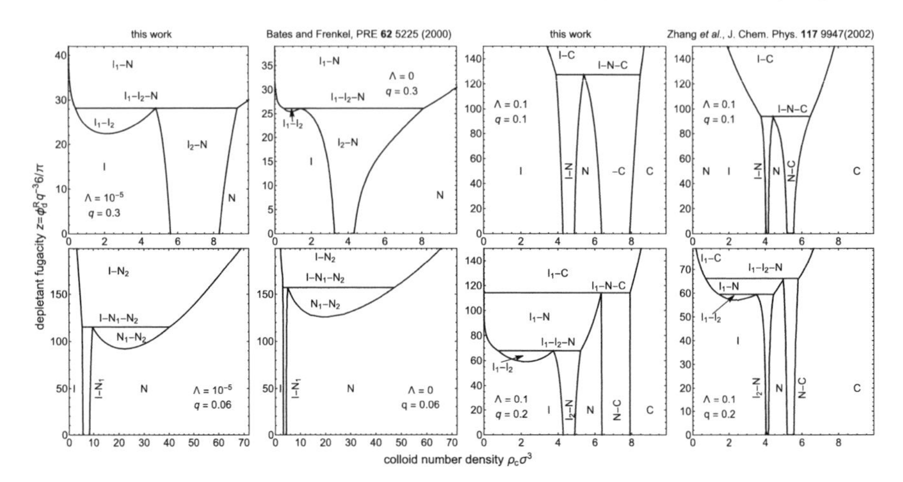

Fig. 6.4 Phase diagrams from FVT compared to results by Bates and Frenkel [17], and compared to the phase diagrams for cut spheres mixed with penetrable hard spheres from Zhang et al. [18]

nematic phase of infinitely thin plates is quite pronounced [26, 29], the depletant-free baselines ($\bar{z} = 0$) deviate from each other. However, the phase sequences occurring at each range of depletion attraction are not affected by the choice of the orientational distribution probability for the nematic phase. We also compare our theoretical approach with the phase diagrams generated by mixtures of thick platelets (described as cut-spheres) plus depletants by Zhang et al. [18] for $\Lambda = 0.1$ and $q = \{0.1, 0.2\}$ (Fig. 6.4). The hybrid MC-FVT approach followed by Zhang et al. [18] is similar to the one of Bates and Frenkel [17]. Even though the free volume fractions (α, Fig. 6.3) are quite close, the depletant fugacity that leads to phase coexistence is higher for our colloidal discs than for these cut-spheres. This is (most likely) due to the higher excluded volume between discs as between cut-spheres, and due to the different mesogen shape used. The overall topology of the phase diagram from FVT agrees with the one obtained from the hybrid method even though the platelet shape differs.

6.3 Multi-phase Coexistences and Critical End Point

In a large part of the $\{\Lambda, q\}$ parameter space we find phase diagrams such as those in Fig. 6.4. Isostructural phase transitions (I_1–I_2, N_1–N_2, and C_1–C_2) lead to a considerable enrichment of the phase diagrams including six three-phase coexistences: I_1–I_2–N, I_1–I_2–C, I–N_1–N_2, N_1–N_2–C, I–C_1–C_2, and N–C_1–C_2; and three four-phase regions: I_1–I_2–N–C, I–N_1–N_2–C, and I–N–C_1–C_2. In this Section we delineate the regions in the $\{\Lambda, q\}$ plane where three- and four-phase regions occur and discuss why these appear. To gain insight into the different types of multi-phase coexistence regions involving isostructural coexistence it is useful to focus on the isostructural phase coexistences in the vicinity of their corresponding critical endpoints [30]. The critical endpoint (CEP) is defined as the condition under which the critical point of the isostructural phases is in equilibrium with a distinct third phase [31].

6.3.1 Isostructural Isotropic Coexistence

We first consider the I_1–I_2 coexistence and check under which conditions it coexists with other phases. The q-value at which the critical point of the I_1–I_2 coexistence meets the I_1–I_2–N three phase region marks the relative range of the depletion interaction below which I_1–I_2 transition becomes metastable with respect to the I–N transition. Above this q value we have a stable three-phase I_1–I_2–N region. Similarly, the q-value at which the critical point of the I_1–I_2 region meets the I_1–I_2–C three phase region marks the relative range of the depletion interaction below which I_1–I_2 transition becomes metastable with respect to the I–C transition; above this q value three-phase I_1–I_2–C equilibria appear. Due to the decrease of the platelet–platelet excluded volume with increasing Λ, the range of the depletion attraction required for an I_1–I_2–N multi-phase coexistence lowers as the discs become thicker. However, when the columnar state dominates over the nematic one at sufficiently high Λ and q, the scenario changes and the I_1–I_2 coexistence is connected to a I_1–I_2–C triple point.

The calculated I_1–I_2–N and I_1–I_2–C critical end points (CEPs, shown in Fig. 6.8) coincide at $\{\Lambda, q\} \approx \{0.122, 0.163\}$, leading to a remarkable (I_1I_2)–N–C CEP where the CEP of the I_1–I_2 critical point is in equilibrium with *two* distinct phases: N and C. For $\Lambda = 0.15$ the transitions from I_1–I_2–C ($q = 0.185$) to I_1–I_2–N ($q = 0.25$) can be observed in Fig. 6.5 (upper panels), where FVT binodals are plotted for various q values. For $q = 0.185$, stable I_1–I_2 coexistence is possible, and an I_1–I_2–C triple line can be observed above the I_2–N–C triple line. For $q = 0.25$ an I_1–I_2–N triple line appears at lower ϕ_d^R than the I_1–N–C triple line. Strikingly, at $q = 0.215$ a I_1–I_2–N–C *four phase* coexistence is predicted: at this q-value the I_1–I_2–N, I_1–N–C, I_1–I_2–C and I_2–N–C three phase lines merge. This I_1–I_2–N–C four-phase coexistence occurs along the curve of $\{\Lambda, q\}$-values from the (I_1I_2)–N–C CEP($\{\Lambda, q\} \approx \{0.122, 0.163\}$) towards the triple-line for platelets in the absence of depletants ($\Lambda \approx 0.16$). No isostructural isotropic coexistence occurs for $\Lambda = 0.15$ if $q \lesssim 0.17$. In the system representation (bottom panels of Fig. 6.5), the area of the multi-phase coexistence denotes the region where they are predicted in terms of colloid and depletion concentrations for the particular set of relevant size ratios.

6.3.2 Isostructural Nematic Coexistence

We now pay attention to the N_1–N_2 coexistence and check under which conditions it coexists with other phases. Again we first consider the N_1–N_2 CEPs: I–N_1–N_2

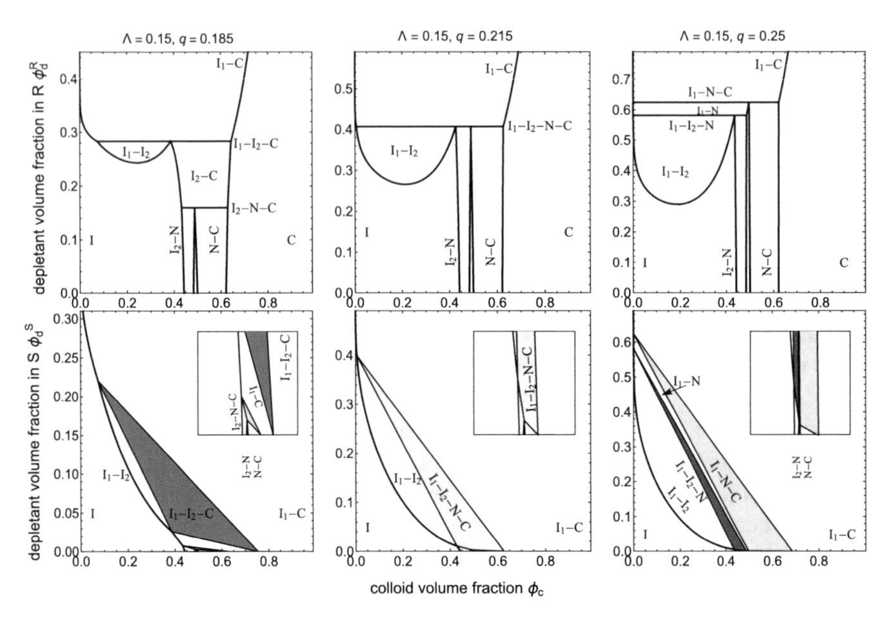

Fig. 6.5 Collection of phase diagrams for platelet-polymer mixtures in the $\{\phi_c, \phi_d^R\}$ phase space for $\Lambda = 0.15$ and various q-values as indicated. Horizontal lines mark multiple-phase coexistence (more than two-phase). Bottom panels: as top ones but in the $\{\phi_c, \phi_d^S\}$ phase space. The coloured triangles are triple regions; they indicate the system representation of the triple-point lines of the top panels. Inset plots of bottom panels zoom into the low-depletant concentration regime

and N_1–N_2–C. From previous calculations [22] it is known that N_1–N_2 equilibria are only stable at low values of Λ and q. In Fig. 6.6 we illustrate the influence of varying q at $\Lambda = 0.02$. For $q = 0.03$ a stable N_1–N_2 coexistence is possible and a N_1–N_2–C triple line emerges, while for $q = 0.04$ an I–N_1–N_2 triple line is present. For $\Lambda = 0.02$ and $q = 0.033$ the N_1–N_2–C, I–N_1–C, I–N_1–N_2 and I–N_2–C three-phase lines merge and again a four phase equilibrium is predicted: a quadruple I–N_1–N_2–C coexistence. In the bottom panels of Fig. 6.6, the areas denote the regions in the system representation where multi-phase coexistences are predicted in terms of colloid and depletant volume fractions for the particular set of size ratios. At fixed $\Lambda = 0.02$, the N_1–N_2 coexistence is metastable with respect to the N–C transition for $q \lesssim 0.03$, becomes stable for $0.03 \lesssim q \lesssim 0.07$, and gets metastable with respect to the I–N for $q \gtrsim 0.07$. The CEP of the four-phase coexistence I–N_1–N_2–C separates the regions where I–N_1–N_2 and N_1–N_2–C three-phase coexistence occur for all Λ (see Fig. 6.8). From these CEPs of the isostructural N_1–N_2 coexistence, a re-entrant behaviour is revealed at fixed Λ.

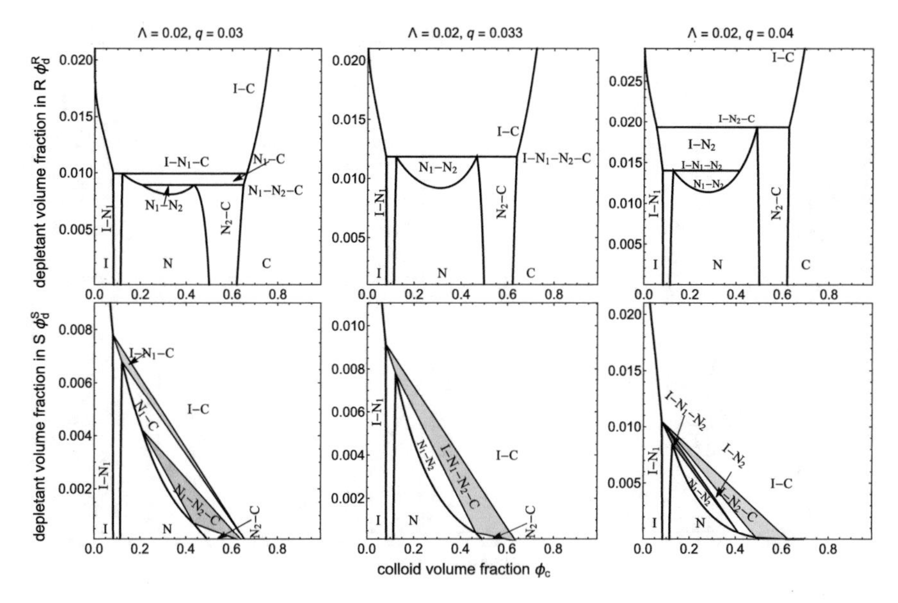

Fig. 6.6 Similar plots as in Fig. 6.5, but for very *thin* hard platelets ($\Lambda = 0.02$) and small polymers, in the range $0.03 \leq q \leq 0.04$. An isostructural nematic–nematic coexistence occurs only within a certain range of depletant sizes

6.3.3 Isostructural Columnar Coexistence: A First Account

Surprisingly, systematically scanning the possible phase equilibria also revealed isostructural columnar phase state coexistence regions. Therefore the focus now is on C_1–C_2 coexistences, and we investigate under which conditions isostructural columnar equilibria coexists with other phases. In Fig. 6.7 we present phase diagrams for $\Lambda = 0.15$, but for small q-values. For $q = 0.01$ a N–C_1–C_2 triple line emerges while for $q = 0.025$ an I–C_1–C_2 triple line is present. For $q = 0.021$ the N–C_1–C_2, I–N–C_1, I–C_1–C_2 and the I–N–C_2 three-phase lines merge and again a four-phase equilibrium is predicted: the I–N–C_1–C_2 coexistence. This four-phase coexistence occurs along a curve of $\{\Lambda, q\}$-values from the I–N–(C_1C_2) CEP at $\{\Lambda, q\} \approx \{0.135, 0.029\}$ towards $\{\Lambda, q\} \approx \{0.155, 0\}$ (see Fig. 6.8). In the system representation (bottom panels of Fig. 6.7), the area of multi-phase coexistence denotes the region where they are predicted in terms of colloid and depletion concentrations for the particular set of relevant size ratios. Isostructural C_1–C_2 coexistence have not been reported yet, higher order multi-phase coexistences for this isostructural coexistence are unknown as well. In light of our results in Chap. 3, a better account of the free volume for depletants in the columnar phase improves the phase diagrams containing this remarkable isostructural C_1–C_2 coexistence. In Sect. 6.4, we focus on the N–C_1–C_2 coexistence and study the effect of considering a more accurate depletant partitioning over the two columnar phases. Note, however, that the simple scaled particle theory (SPT) expression for α_C allowed us to systematically vary the system parameters, and revealed

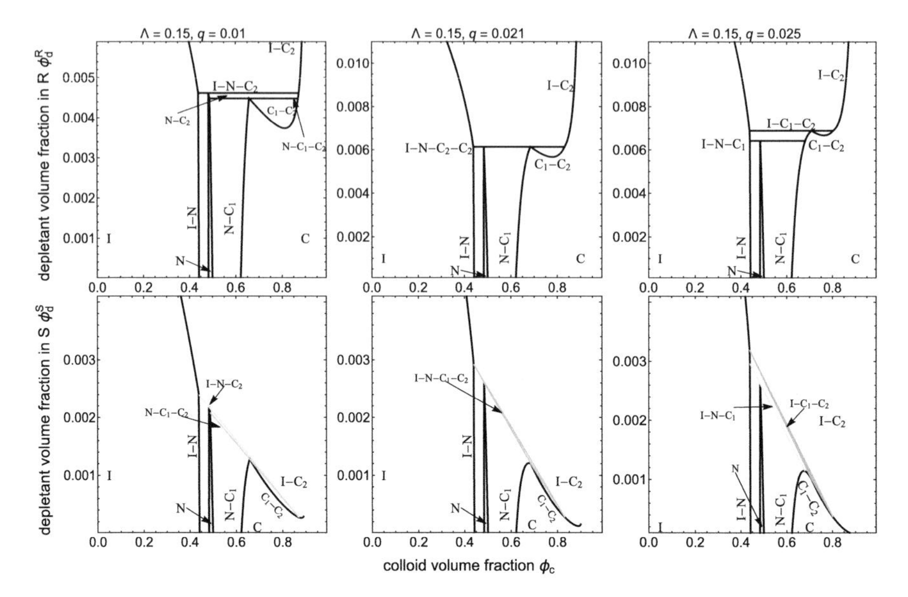

Fig. 6.7 Similar plots as in Fig. 6.5, but for *thick* hard platelets ($\Lambda = 0.15$) and tiny polymers, in the range $0.01 \leq q \leq 0.025$. For thick platelets and tiny polymers, a remarkable isostructural columnar–columnar coexistence occurs (analysed in details in Sect. 6.4)

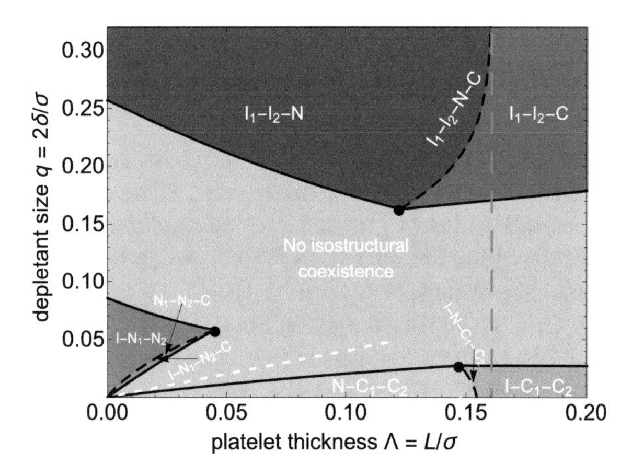

Fig. 6.8 Generic overview of isostructural multi-phase coexistences found in a colloidal platelet–polymer mixture in terms of the platelet aspect ratio (Λ) and the relative size of depletant (q). Solid curves correspond to critical end point (CEP) curves of the isostructural phase coexistences as indicated. Quadruple coexistence *curves* (dashed) bound the corresponding isostructural coexistence *regions*. The CEPs of the quadruple coexistences are marked by black dots, which indicate the points where *two* isostructural CEP curves coincide. The long-dashed, grey, vertical line corresponds to the triple point (I–N–C) for platelets in the absence of depletants ($\Lambda \approx 0.16$). Note that three-phase coexistence regions correspond to areas in the $\{\Lambda, q\}$ phase space, whereas four-phase coexistences are only found along their respective curves. The tiny region marked with an arrow corresponds to the N_1–N_2–C coexistence. The short-dashed white curve marks the CEP of N–(C_1C_2) calculated with the modified theory presented in Sect. 6.4

for the first time the C_1–C_2 coexistence. Results in Sect. 6.4 are more accurate, yet also less tractable than the SPT prediction [Eq. (6.5)] due to the relative complexity of the geometrically-based α_C.

6.3.4 *Multiphase Coexistence: Overview and Experimental Comparison*

We calculated the possible thermodynamically stable phases via the CEPs of the different isostructural critical points in equilibrium with a distinct third phase. The results are summarised in Fig. 6.8. In the infinitely thin platelet limit ($\Lambda \to 0$), the stable phases found correspond to those of Bates and Frenkel [17]. The trends with increasing platelet thickness correspond to hybrid theory/simulation approaches for cut-spheres mixed with PHSs [18, 19]. The area covered by the I–N_1–N_2 triple equilibrium matches with previous theoretical studies [22], and reveals possible new regions for multi-phase coexistence. At $\{\Lambda, q\} \approx \{0.045, 0.058\}$, the I–($N_1N_2$) and ($N_1N_2$)–C CEP lines coincide, hence defining the I–(N_1N_2)–C CEP leading to a small nose-like region in which the N_1–N_2 transition is stable (see Fig. 6.8). This region is in agreement with the results reported by Aliabadi et al. [22] using a canonical Onsager–Parsons-Lee approach for colloidal discs-hard sphere mixtures to obtain the boundaries of the stable isostructural N_1–N_2 transition regions. The I–N_1–N_2–C four-phase coexistence occurs along a curve of $\{\Lambda, q\}$-values from the I–(N_1N_2)–C CEP at $\{\Lambda, q\} \approx \{0.045, 0.058\}$ towards the corner of the $\{\Lambda, q\}$ diagram. This quadruple curve practically coincides with the (N_1N_2)–C CEP curve (see Fig. 6.8). Hence, the area covered by the N_1–N_2–C three-phase coexistence is very small.

The isostructural C_1–C_2 coexistence takes place at very low q-values (as shown in Fig. 6.8). This is reminiscent of the occurrence of two isostructural solid phases in a system of spheres with a very narrow range attraction [32, 33]. Again we first consider the C_1–C_2 CEPs: I–(C_1C_2) and N–(C_1C_2). With increasing Λ and at low q-values the N_1–N_2 transition becomes metastable with respect to the N–C transition. On further lowering q we obtain the N–(C_1C_2) CEP (see Fig. 6.8). With further increasing Λ the isotropic state takes over from the nematic in this competition, and we obtain the I–(C_1C_2) CEP (see Fig. 6.8). At $\Lambda = 0.135$ and $q = 0.025$ the I–(C_1C_2) and N–(C_1C_2) CEP curves coincide, and we obtain the I–N–(C_1C_2) CEP. The increase in possible q-values leading to C_1–C_2 with increasing Λ follows from the fact that the free volume for depletants in the direction perpendicular to the column (in the interstices of the columns) increases with increasing Λ. This becomes clear upon close inspection of the free volume fraction for depletants in the columnar state. Therefore, we pay close attention to the isostructural N–C_1–C_2 coexistence and its critical end behaviour in Sect. 6.4.

The overview shown in Fig. 6.8 constitutes the main result of this research: it provides a systematic overview of where the possible critical points, three-, and four-phase coexistence areas occur in terms of the (geometrical) system parameters $\{\Lambda, q\}$.

It may be questioned whether four-phase coexistences violate the Gibbs phase rule, which dictates that in a system of $\mathfrak{C}$ components the maximum number of phases that can coexist is $\mathfrak{C} + 2$. As shown by Tanaka and co-workers [34], for systems with additional model parameters, four phases can coexist even in single component systems for specific values of these parameters. In the system considered here there are only hard interactions between the two components (solvent is considered as background) so the temperature is effectively not a degree of freedom. Thus the maximum number of phases that can coexist is $\mathfrak{C} + 1$. So at first sight with $\mathfrak{C} = 2$ a four phase equilibrium does not seem possible. However, here we have two additional parameters at our disposal, Λ and q, which act as additional field variables, thus allowing more than $\mathfrak{C} + 1$ phases to be simultaneously in equilibrium, at appropriately constrained values of these parameters. Figure 6.8 shows that for this particular system it is not possible to find coexistence between more than four phases. For other particle shapes (e.g., biaxial plates providing an additional system parameter) it may be possible to find a quintuple or even a sextuple point. A mixture showing two simultaneous isostructural transitions (e.g., N_1–N_2 and C_1–C_2) could perhaps lead to such conditions. Alternatively, systems that exhibit more possible phase states could potentially lead to such depletion-driven multi-phase coexistences [35].

In experiments where colloidal platelets, which in the pure state give rise to liquid crystal phases, are mixed with nonadsorbing polymers [5–7] or with small spheres [8–15] the phase behaviour changes significantly compared to that of the system of pure platelets. We have to realise that the depletion interaction in combination with the inherent polydisperse nature of the colloids [5, 36, 37] and the sedimentation equilibrium under gravity [38, 39] may give rise to the presence of additional multi-phase coexistences. For example Wensink and Lekkerkerker [38] showed that a four phase equilibrium I_1–I_2–N–C may arise in the gravitational field even if the system 'without gravity' only displays the two phase equilibria I_1–I_2, I_2–N, N–C. De las Heras et al. [13, 39] showed that in a system with only one isotropic phase, two isotropic layers may appear with a nematic phase floating in between (in a system where gravity does not play a role). Moreover, most experimental colloidal systems are non-hard and contain additional direct interactions. Here we briefly consider the experimentally observed multi-phase coexistences for mixtures of colloidal platelets with nonadsorbing polymers.

Van der Kooij et al. [5] studied a system of sterically stabilised Gibbsite platelets dispersed in toluene mixed with added non-adsorbing polymer polydimethylsiloxane (PDMS). In this plate–polymer mixture, the aspect ratio and relative polymer size are $\Lambda \approx 0.07$ and $q \approx 0.35$. The sterically stabilised Gibbsite platelets in their pure state exhibit an I–N transition and a N–C transition with increasing concentration [40]. In the Gibbsite-PDMS mixture a four-phase equilibrium I_1–I_2–N–C, bordered by three three-phase equilibria I_1–I_2–N, I_1–N–C, and I_2–N–C was observed. The results were rationalised by representing the polydisperse platelets by a bidisperse system consisting of platelets of lower Λ and higher Λ. As we have seen in our calculations (see Fig. 6.8) this gives rise to the three-phase regions I_1–I_2–N and I–N–C for the low Λ and the three-phase regions I_1–I_2–C and I–N–C for the high Λ. In the three-dimensional concentration diagram spanned by the concentrations of the lower Λ platelets, the higher Λ platelets and the polymer, the four three-phase regions intersect and a tetrahedron-shaped four-phase I_1–I_2–N–C region appears bordered by four three-phase regions: I_1–I_2–N, I_1-N–C, I_1–I_2–C and I_2–N–C. The projection of these four- and three-phase regions on the experimental plane then leads to the referred observations.

Zhu et al. [6] studied a system of positively charged Mg_2Al layered double hydroxide platelets mixed with polyethylene glycol (PEG). Those platelets exhibit an I–N transition [41]. In this plate–polymer mixture, $\Lambda \approx 0.055$ and $q \approx 0.01$. A multi-phase coexistence consisting of a dilute upper phase, two or three birefringent phases and an amorphous bottom phase was observed. These results are ascribed to a combination of the depletion interaction and sedimentation equilibrium. The nature of the birefringent phases was not determined; however in one experiment a phase which gives Bragg reflections of visible light was observed indicating that there is positional order in this phase. So this phase may be a columnar phase, which given the values of Λ and q in this experiment would agree with the results given in Fig. 6.8.

Luan et al. [7] observed multiphase coexistences consisting of up to six phases in a system of positively charged Mg_2Al layered double hydroxide platelets mixed with polyvinyl pyrrolidone (PVP) with $\Lambda \approx 0.026$ and $q \approx 0.78$: a dilute upper phase, two isotropic phases, two nematic phases, an amorphous bottom phase, and two or three birefringent phases. They ascribe these results to a combination of the depletion interaction, sedimentation equilibria and polydispersity effects.

6.4 Compartmentalisation in Crowded Discotics: Detailed Account of C_1–C_2

In this section we study the stability of the columnar–columnar coexistence via direct coexistence Monte Carlo (MC) computer simulations (see Sect. 6.C), and compare it with a modified FVT where the depletant partitioning in the columnar phase is more accurately accounted for than in the previous Sections. To retain a nematic–columnar (N–C) depletant-free coexistence closer to simulation results (Fig. 6.9, triangles), we consider a corrected free energy for the columnar phase [29, 42]:

$$\widetilde{F}_{\mathrm{C}}^{\mathrm{alt}} = \widetilde{F}_{\mathrm{C}} - 1.386\,. \tag{6.8}$$

Crucially, we also consider a geometrical free volume fraction for depletants in the columnar phase, $\alpha_{\mathrm{C}}^{\mathrm{geo}}$ (see Sect. 6.B). The predicted phase diagram for $\Lambda = 0.1$ and $q = 0.01$ is presented in the left panel of Fig. 6.9 (black curves), and compared to that obtained using $\widetilde{F}_{\mathrm{C}}^{\mathrm{alt}}$ and the scaled particle theory (SPT) expression for α_{C} (dashed grey curves). As expected from Chap. 3, a more accurate account of the depletant partitioning over the columnar phases affects the *entire* phase diagram: all phase equilibria shift to lower depletant concentrations. The I–C_2 coexistence attains (much) more phase space than with the SPT-derived α_{C}. We note that the

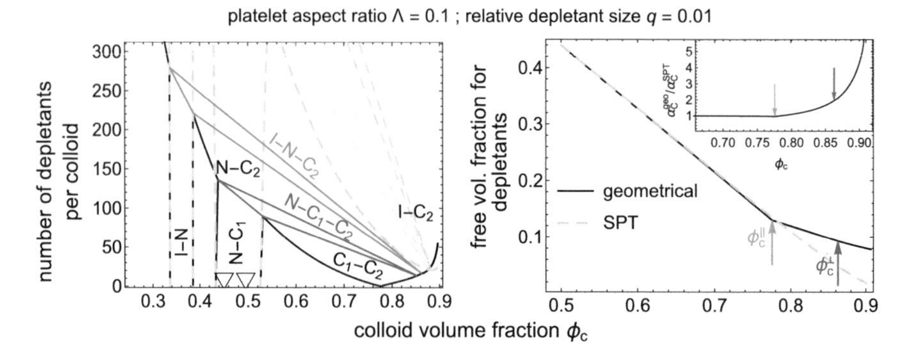

Fig. 6.9 Left panel: Phase diagrams of discotic colloids mixed with PHSs using the geometrical free volume fraction [Eq. (6.26)] (black curves) or the fluid-like free volume fraction [Eq. (6.5)] (dashed grey curves) for PHSs in the columnar phase. Down-triangles are independent simulation results [27]; bounded areas indicate three-phase coexistences. Coexistences indicated are only for the geometrical free volume fraction results. Right panel: Free volume fraction of depletants in the columnar phase. Inset: ratio of Eq. (6.26) to Eq. (6.5)

phase sequences occurring are as predicted with the SPT expression for α_C (in line with Sect. 6.2.3). We use, for the phase diagrams in this Section, the number of depletants per colloidal particle: $N_d^S/N_c \equiv v_c\phi_d^S/(v_d\phi_c)$.

We compare the two different α_C in the right panel of Fig. 6.9. From geometrical considerations, it is clear that the the SPT-prediction *underestimates* α_C at high colloid volume fractions ϕ_c. For these specific $\{\Lambda, q\}$-values, the colloid volume fraction at which overlap of the depletion zones in the direction parallel to the column $\phi_c^{\|}$ occurs is indicated. We denote this direction as intra-columnar, $r^{\|}$. This $\phi_c^{\|}$-value marks the onset of deviation between the SPT and the geometrical α_C (see inset). Furthermore, the C_1–C_2 critical point occurs precisely at $\phi_c^{\|}$. These two results apply for all $\{\Lambda, q\}$ with $q \lesssim 0.05$ and $\Lambda \lesssim 0.12$, and follow from our α_C^{geo} expression. According to this α_C^{geo}, in the lower-density columnar phase (C_1) depletants are present between the flat faces of the discotics [see Fig. 6.10(e)], as opposed to the denser phase (C_2) [see Fig. 6.10(f)]. Consequently, the C_1–C_2 coexistence is driven by the depletant partitioning in $r_{\|}$. Note that there is always space for depletants in the interstices between columns (the pockets of the crowded colloidal state). Overlap of depletion zones in the direction perpendicular to the column (inter-columnar direction, $r_\perp$) has a barely perceptible effect on α_C, and occurs from $\phi_c^\perp$ (marked with an arrow in Fig. 6.9). Overlap of depletion zones occurs at lower colloid volume fraction in $r_{\|}$ than in $r_\perp$, and follows from the different scaling of the intra- and inter-columnar plate–plate spacings with ϕ_c [43].

Isostructural dense phase coexistences have been obtained for hard spheres (HSs) interacting via short-ranged (direct) attractions (namely, isostructural face centred cubic coexistence) [32, 44, 45]. In contrast to these approaches, we consider here *anisotropic* particles with excluded volume interactions not only among themselves, but also with a second component in the mixture. The C_1–C_2 coexistence is fully

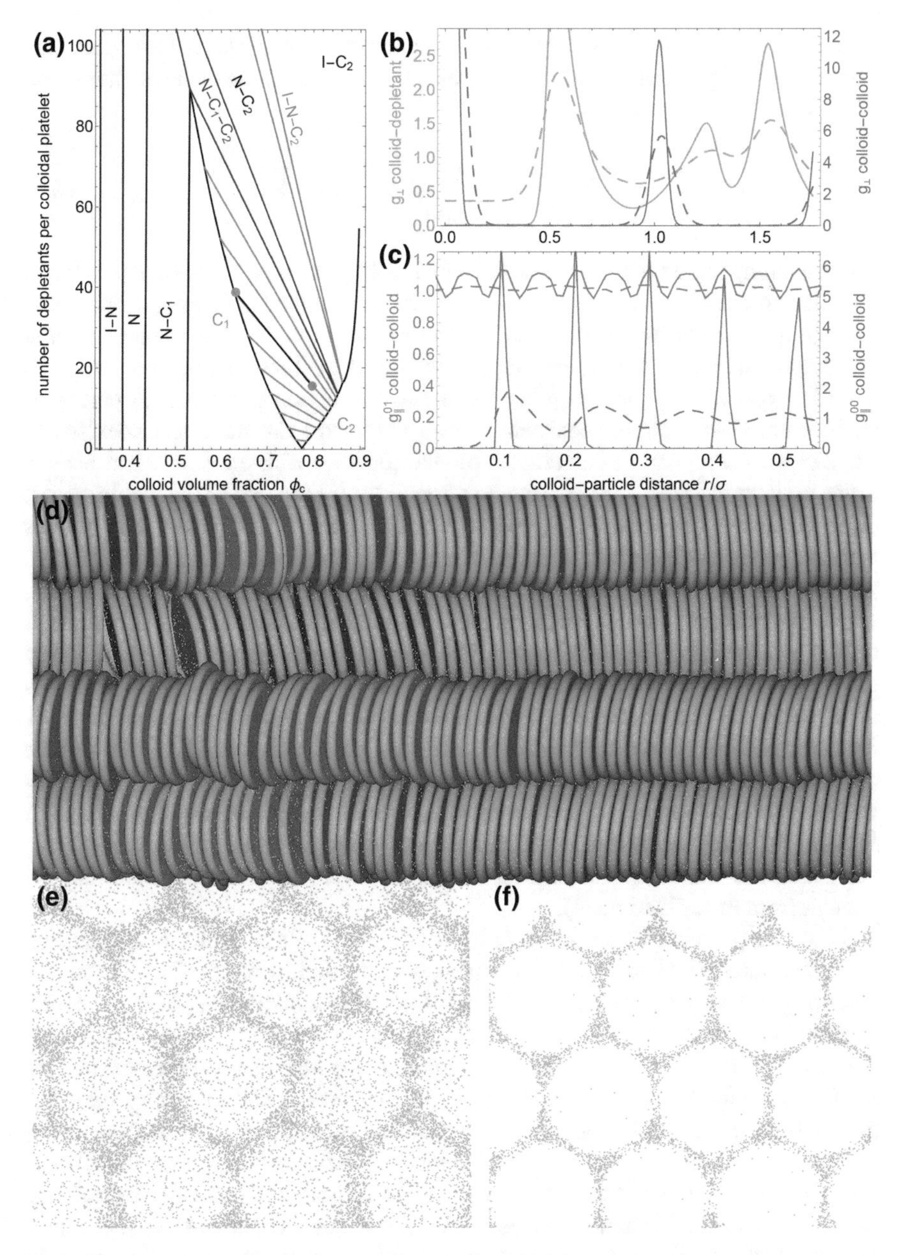

Fig. 6.10 **a** Zoom in from the phase diagram in Fig. 6.9 around the vicinity of the C_1–C_2 coexistence (illustrative tie-lines in orange), considering only the geometrical free volume fraction for depletants in the columnar phase. Orange circles correspond to the equilibrium simulation data, tie-line in black. **b, c** Distribution functions in the inter-columnar **d** [$g_\perp$] and intra-columnar **e** [$g_\parallel$] directions for the particle-pairs indicated; the difference between $g_\parallel^{00}$ and $g_\parallel^{01}$ is explained in main text. **d–f** Simulation snapshots of colloids (brown) and depletants (green) in equilibrium states. (d) Snapshot from the (1100) plane of a final direct-coexistence simulation, where the C_1 phase occurs on the left and the C_2 on the right of the simulation box. **e, f** Snapshots from the (0001) plane of the depletants present in the equilibrium lower-density [C_1, (e)] and higher-density [C_2, (f)] columnar phases

understood (and predicted) solely in terms of compartmentalisation of the depletants over the two (highly dense) phases. It is noted that the maximum depletion attraction $W_{\text{AOV}}^{\text{max}}$ between hard platelets

$$\frac{W_{\text{AOV}}^{\text{max}}}{k_B T} = -\phi_{\text{d}}^{\text{R}} \left(\frac{3}{4q^2} + \frac{3\pi}{8q} + 1 \right) \tag{6.9}$$

is much stronger than between HSs (for which $W_{\text{AOV}}^{\text{max}} \propto -\phi_{\text{d}}^{\text{R}} q^{-1}$, see Chap. 3) when considering addition of tiny depletants. The tendency of the flat faces of the discotics to align at high concentrations is enhanced by the presence of the non-adsorbing depletants, leading to a $W_{\text{AOV}}^{\text{max}} \propto -\phi_{\text{d}}^{\text{R}} q^{-2}$ scaling in the limit of small q.

In Fig. 6.10a, we zoom in on the phase-diagram region of interest. A particular equilibrium C_1–C_2 phase coexistence result obtained from the direct coexistence Monte Carlo computer simulations is plotted together with the theoretical results. Some simulation details are presented in Sect. 6.C. A snapshot of the (equilibrated) direct coexistence is shown in Fig. 6.10d. For computational purposes, the discotics were modelled as oblate hard spherocylinders (OHSCs) instead of as hard cylinders [28]. The close-packing fraction for these OHSCs ($\Lambda = 0.1$) is $\phi_{\text{c}}^{\text{cp}} \approx 0.88$ [28], which explains the lower ϕ_{c}-value on the C_2 branch of the simulations as compared to FVT predictions. The stacking of OHSCs in the columnar phase is also slightly different than those of hard cylinders due to their rounded edges [28]. Besides this offset in the C_2 branch, the direct coexistence MC simulation results and the FVT predicted tie-line are in remarkable agreement. As previously noted, snapshots of the different (independently equilibrated) plate–polymer mixtures [Fig. 6.10e, f] show that the depletant compartmentalisation is in line with the predictions from the $\alpha_{\text{C}}^{\text{geo}}$. Further, the direct coexistence MC simulations show that this isostructural C_1–C_2 coexistence is stable against fluctuations (at least for depletant concentrations far enough from the critical point).

Next, we pay attention to the colloid–colloid and colloid–depletant distribution functions obtained from the MC simulations, presented in Fig. 6.10b, c (solid and dashed green curves). The most insightful of these distributions is the colloid–depletant distribution function in $r_\perp$ [$g_\perp^{\text{c-d}}$, green curves in Fig. 6.10b]. For the C_2 phase, $g_\perp^{\text{c-d}} \approx 0$ for $r_\perp \lesssim 0.5\sigma$, which confirms that there are barely depletants present in the intra-columnar direction in C_2 (solid green curves). On the contrary, there is a clearly homogeneous distribution of depletants on the top and bottom of the mesogenic flat faces in the C_1 phase: $g_\perp^{\text{c-d}} \approx 0.4$ for $r_\perp \lesssim 0.5\sigma$ (dashed green curves). The first peak at $r_\perp \approx 0.5\sigma$ of this $g_\perp^{\text{c-d}}$, which is present both in C_1 and C_2, corresponds to the doughnut-like volume available for depletants between discotics. The similar position of the second and third peak of $g_\perp^{\text{c-d}}$ indicate that depletants are present in the interstices of both columnar phases. Furthermore, the $g_\perp$-value at these peaks is significantly higher in the C_2 phase than in the C_1 one. This follows naturally from the lack of pockets in $r_\parallel$ in C_2, which leads to accumulation of the depletants in the interstices. In the inter-columnar direction $r_\perp$, the colloid–colloid distribution $g_\perp^{\text{c-c}}$ shows peaks corresponding to the hexagonal (two-dimensional) arrangement both for

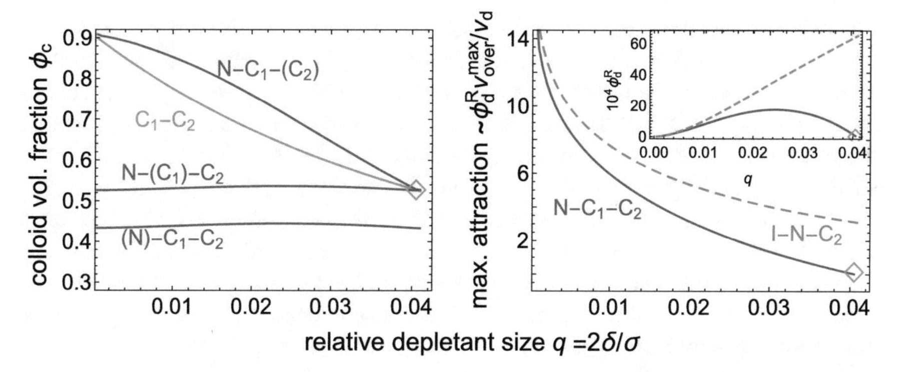

Fig. 6.11 Left panel: colloid volume fraction ϕ_c at the nematic–columnar–columnar N–C$_1$–C$_2$ triple point (dark grey curves) and at the columnar–columnar (C$_1$–C$_2$) critical point (light grey curve) as a function of the relative depletant size q. For the three phase coexistence, the parenthesis denotes the phase referred to. Right panel: maximum strength of the depletion attraction at N–C$_1$–C$_2$ (dark grey curve) and at the I–N–C$_2$ triple points (dashed grey curve) as a function of the relative depletant size q. In the inset, the depletant concentration in the reservoir is shown. The open symbols denote the critical endpoint

C$_1$ and C$_2$ (brown dashed and solid curves). In contrast, colloid–colloid distributions in the intra-columnar direction $r_\parallel$, both within the same column $g_\parallel^{00,\text{c-c}}$ and between different columns $g_\parallel^{01,\text{c-c}}$, manifest a solid-like behaviour of the C$_2$-phase (brown solid curves) and a more fluid-like behaviour of the C$_1$-phase in $r_\parallel$ (dashed brown curve)[see Fig. 6.10c]. We deduce from the colloid–colloid and colloid–polymer distributions that: (i) The C$_1$–C$_2$ coexistence is solid–solid like for the discotics in $r_\perp$ but solid–fluid like in $r_\parallel$; and (ii) Depletants compartmentalise according to the pockets present in the columnar phase. In the C$_1$ phase, pockets are available both in $r_\parallel$ and $r_\perp$. Opposite to this, in the C$_2$ phase, pockets are only available the interstices (i.e., in $r_\perp$).

We finally turn our attention to the N–(C$_1$C$_1$) critical end point (CEP) for which FVT predictions are plotted in Fig. 6.11. As observed for HSs mixed with PHSs at low q (Chap. 3), ϕ_c at the C$_1$–C$_2$ critical point ($\phi_c^\parallel$, light grey curve) decreases with increasing q. The N–C$_1$–C$_2$ CEP occurs at $q \approx 0.04$, significantly above the original prediction ($q \approx 0.02$, see Sect. 6.3.4). The depletant concentration at the CEP is presented in the right panel of Fig. 6.11. The depletant concentration of at the I–N–C$_2$ coexistence is always above the one of the N–C$_1$–C$_2$. In line with our results in Chap. 3, the PHS volume fraction at the N–C$_1$–C$_2$ triple point first increases and then decreases with increasing q (inset of Fig. 6.11). This denotes a soft re-entrant behaviour [45]. However, the maximum strength of the depletion attraction, Eq. (6.9), decays at the N–C$_1$–C$_2$ triple point. The C$_1$–C$_2$ critical point occurs at a depletant concentration which is virtually zero: it may appear that the mere presence of depletants in the system is enough to induce two different columnar states. While not reported previously for colloidal systems (as far as we are aware of), such behaviour of the critical point is quite common in alloys [46]. One may argue that for such tiny depletants the columnar phase contains *two effective* different systems: one in $r_{\parallel}$ and one in $r_{\perp}$. The depletant–colloid distributions obtained from the MC simulations support the idea of these two different subsystems for tiny depletants in the columnar phases. The thicker the discotic, the more clear the difference between these 'two effective systems' is. This may explain the increase in q at the N–(C_1C_1) CEP with increasing Λ observed in Fig. 6.8. We finally note that the N–(C_1C_1) CEP dependence with the discotic aspect ratio Λ remains as predicted with the more tractable SPT approach originally followed (see white dashed curve in Fig. 6.8).

The insights put forward in this Section shine light on the role of excluded volume interactions in compartmentalisation in crowded and highly size-asymmetric environments, which are of relevance in biological systems [47, 48].

6.5 Conclusions

Free volume theory (FVT) is a versatile and tractable framework to predict the phase behaviour of mixtures of platelets and non-adsorbing polymer chains in a common solvent. We reveal via FVT, a multi-phase coexistence overview in terms of the platelet thickness (Λ) and the relative depletant size (q) was obtained (Fig. 6.8). The possible phase states of the canonical platelet system were considered: isotropic (I), nematic (N) and columnar (C). The final phase diagrams do not only match with previous theoretical approaches and with experimental results but also exhibit a columnar-columnar isostructural coexistence not reported before. On top of a I_1–I_2–N–C quadruple coexistence, two other four-phase coexistences are presented involving orientationally ordered isostructural coexistences at low depletant sizes: I–N_1–N_2–C, and I–N–C_1–C_2. All quadruple coexistences arise when two different isostructural triple phase coexistences merge. The stability regions can be explained in terms of excluded volume repulsions between hard discs being reduced by the second component in the mixture. The appearance of columnar phases can be rationalised in terms of alignment (and stacking) of the flat faces of the colloidal hard discs, which increases the free volume and the entropy of the depletants.

The isostructural phase coexistences of ordered phase states of the discs (N_1–N_2 and C_1–C_2) are driven by short-ranged attractions (small q and low depletant concentrations). This may be envisaged as an effective 'sticky hard platelet' interaction, which is supported by the presence of columnar phase equilibria (C_1–C_2, I–C_1–C_2, N–C_1–C_2, I–N–C_1–C_2) for relatively small depletants. On the other hand, the isotropic isostructural coexistence is driven by a large depletion zone that sufficiently smooths the interacting platelet volume. Hence, the relative size of the depletant modifies the coexistence landscape, enhancing isostructural coexistences between partially crystalline phases (C_1–C_2 and N_1–N_2) for small depletant sizes while promoting isotropic fluid-fluid (I_1–I_2) coexistence for large enough depletants.

The theoretically predicted C_1–C_2 coexistence, driven by the depletant partitioning in the intra-columnar direction, was confirmed by comparison with direct coexistence Monte Carlo simulations explicitly accounting for a binary mixture of discotics and depletants. It follows both from geometrical considerations and Monte Carlo simulations that there are two preferred pockets for tiny depletants in a columnar phase; one in the intra-columnar and the other inter-columnar direction. Via a geometrical free volume fraction for depletants, the role of excluded volume interactions in a crowded and highly asymmetric system can be isolated.

6.A Thermodynamics of Pure Platelet Suspensions

Various thermodynamic properties of pure platelet suspensions have been studied in detail previously [26, 49], and we solely report here the key ingredients required to calculate the final phase diagrams of model colloidal platelet-polymer mixtures. Entropy-driven phase transitions [50] as considered here depend on the excluded volume between two colloidal particles. This excluded volume is defined as the volume inaccessible to a second particle in the system as a consequence of the presence of a first particle. For two colloidal platelets, the excluded volume ($v_{\rm exc}^{\rm p\text{-}p}$) per particle volume ($v_{\rm c}$) reads [51]:

$$\frac{v_{\rm exc}^{\rm p\text{-}p}}{v_c} = 2\left[|\cos\gamma| + \frac{4E\,(\sin\gamma)}{\pi} + 1\right] + \frac{8\Lambda\sin\gamma}{\pi} + \frac{2\sin\gamma}{\Lambda}, \tag{6.10}$$

where γ defines the relative orientation between two colloidal discs, and $E(x)$ is the complete elliptic integral of the second kind. Considering the symmetry of a platelet, its orientation can be defined via a unit vector ($\hat{u}$) in the axis of symmetry of the cylinder. Hence, $\cos\gamma = \hat{u}\cdot\hat{u}'$, where $\hat{u}'$ simple refers to a second cylinder. For the isotropic and nematic phases we consider Onsager–Parsons-Lee theory [26, 52]. The free energy of both the isotropic and nematic phases reads:

$$\frac{\widetilde{F}_k}{\phi_{\rm c}} = \ln\widetilde{v}_{\rm c} + \ln\phi_{\rm c} - 1 + \sigma_k[f(\hat{u})] + \frac{2}{\pi}\frac{\phi_{\rm c}}{\Lambda}G_{\rm P}\langle\langle\Theta_{\rm Exc}(\hat{u},\hat{u}')\rangle\rangle. \tag{6.11}$$

The first two terms on the right-hand-side of Eq. (6.11) correspond to the finite-volume normalization of the energy ($\widetilde{v}_{\rm c}$ being the dimensionless thermal volume of a platelet) and the ideal gas contribution to the free energy.

In order to calculate the free energy of the isotropic and nematic phases, an orientational distribution function [ODF, $f(\hat{u})$] needs to be accounted for. A system in which particle orientations are taken into account can be envisaged as a multi-component system in which each component corresponds to a possible particle orientation [51, 53]. Hence, the ODF is a measure of the probability of finding a particle with a given orientation $\hat{u}$. The rotational entropy term, $\sigma_k[f(\hat{u})]$ is defined as [52]:

$$\sigma_k[f(\hat{u})] = \int f(\hat{u})\ln[4\pi f(\hat{u})]{\rm d}\hat{u}. \tag{6.12}$$

The dimensionless ensemble-averaged excluded volume follows from:

$$\langle\langle \Theta_k^{\mathrm{Exc}}(\hat{u}, \hat{u}')\rangle\rangle = \frac{1}{\sigma^3}\int\int f(\hat{u}) f(\hat{u}') v_{\mathrm{exc}}^{\mathrm{p\text{-}p}}(\hat{u}\cdot\hat{u}')\mathrm{d}\hat{u}\mathrm{d}\hat{u}'. \tag{6.13}$$

Finally, effects beyond the second osmotic virial coefficient are accounted for in an approximate manner via the Parsons-Lee scaling factor [54, 55]:

$$G_{\mathrm{P}} = \frac{4 - 3\phi_{\mathrm{c}}}{4(1-\phi_{\mathrm{c}})^2}.$$

Formally, at each platelet concentration the free energy of the system must be minimized with respect to the ODF, $f(\hat{u})$. Analytical expressions for the ODF can be obtained for the isotropic state by considering equiprobability of orientations: $f(\hat{u}) = 1/(4\pi)$. In this case, $\sigma_{\mathrm{I}}[f(\hat{u})] = 0$, and by applying the so-called isotropic averages ($\langle\langle \sin\gamma\rangle\rangle_{\mathrm{I}} = \pi/4$, $\langle\langle E\{\sin\gamma\}\rangle\rangle_{\mathrm{I}} = \pi^2/8$, and $\langle\langle \cos\gamma\rangle\rangle_{\mathrm{I}} = 1/2$), the free energy of an isotropic ensemble of discs can be written as [26]:

$$\frac{\widetilde{F}_{\mathrm{I}}}{\phi_{\mathrm{c}}} = \ln\widetilde{v}_{\mathrm{c}} + \ln\phi_{\mathrm{c}} - 1 + \frac{2}{\pi}\frac{\phi_{\mathrm{c}}}{\Lambda} G_{\mathrm{P}}\Theta_{\mathrm{I}}^{\mathrm{Exc}}, \tag{6.14}$$

with:

$$\Theta_{\mathrm{I}}^{\mathrm{Exc}} = \frac{\pi^2}{8} + \left(\frac{3\pi}{4} + \frac{\pi^2}{4}\right)\Lambda + \frac{\pi\Lambda^2}{2}. \tag{6.15}$$

Furthermore, closed expressions for the free energy of the nematic phase can be obtained via a Gaussian approximation [56] for $f(\hat{u})$. Considering that all relative orientations can be defined as a Gaussian perturbation from the nematic director vector provides a closed-form expression for the free energy [26]. This Gaussian ODF reads

$$f_{\mathrm{N\text{-}G}}(\theta) = \frac{\kappa}{4\pi}\exp\left[-\frac{1}{2}\kappa\theta^2\right], \tag{6.16}$$

where θ is the polar angle between the nematic director and the orientation of the platelet. Minimizing the free energy with respect to the unknown parameter of the Gaussian ODF κ provides a closed form for the free energy of the nematic phase [26]:

$$\frac{\widetilde{F}_{\mathrm{N\text{-}G}}}{\phi_{\mathrm{c}}} = \ln\widetilde{v}_{\mathrm{c}} + \ln\phi_{\mathrm{c}} - 1 + \sigma_{\mathrm{N\text{-}G}} + \frac{2}{\pi}\frac{\phi_{\mathrm{c}}}{\Lambda} G_{\mathrm{P}}\Theta_{\mathrm{N\text{-}G}}^{\mathrm{Exc}}, \tag{6.17}$$

with:

$$\sigma_{\text{N-G}} = \ln\left[\frac{\pi\phi_c^2 G_P^2}{\Lambda^2}\right] - 1,$$

and

$$\Theta_{\text{N-G}}^{\text{Exc}} = \frac{1}{2}\pi^{3/2}\kappa(\phi_c, \Lambda)^{-1/2} + 2\pi\Lambda.$$

For the columnar phase, a modified Lennard-Jones–Devonshire (LJD) cell-theory [57] approach provides a closed expression for the free energy [26, 43]:

$$\frac{\widetilde{F}_C}{\phi_c} = \ln\widetilde{v}_c + \ln\phi_c - 3 - 2\ln\left[1 - \frac{1}{\widehat{\Delta}_\perp}\right] + 2\ln\left[\frac{3\widehat{\Delta}_\perp^2\phi_c^r}{2\Lambda\left(1 - \widehat{\Delta}_\perp^2\phi_c^r\right)}\right] - \ln\left[\frac{1}{3}\left(1 - \widehat{\Delta}_\perp^2\phi_c^r\right)\right], \tag{6.18}$$

where the lateral spacing (inter-columnear direction) is:

$$\widehat{\Delta}_\perp \equiv \Delta_\perp/\sigma = \frac{\sqrt[3]{2}\bar{K}^{2/3} - \sqrt[3]{3}4\phi_c^r}{6^{2/3}\sqrt[3]{\bar{K}}\phi_c^r}, \tag{6.19}$$

where

$$\bar{K} = \sqrt{3(\phi_c^r)^3(243\phi_c^r + 32)} + 27(\phi_c^r)^2, \tag{6.20}$$

and with

$$\phi_c^r = \phi_c/\phi_c^{cp}, \tag{6.21}$$

and

$$\phi_c^{cp} = \pi/(2\sqrt{3}) \approx 0.907. \tag{6.22}$$

The approximated expression for the intra-columnar spacing [43] $\Delta_\parallel$ is:

$$\widehat{\Delta}_\parallel \equiv \Delta_\parallel/L = \frac{1}{\phi_c^r\widehat{\Delta}_\perp^2} \tag{6.23}$$

The (dimensionless) osmotic pressures and chemical potentials follow the relations in Chap. 1:

$$\beta\mu \equiv \widetilde{\mu} = \left(\frac{\partial\widetilde{F}_c}{\partial\phi_c}\right)_{T,V} \quad ; \quad \beta\Pi v_c \equiv \widetilde{\Pi} = \phi_c\widetilde{\mu} - \widetilde{F}_c, \tag{6.24}$$

6.B Geometrical Free Volume Fraction in the Columnar State

We follow the ideas put forward in Chap. 3 to calculate a geometrical free volume fraction for PHS in a columnar state. Let V_{UC} be the volume of the columnar unit cell, such that:

$$V_{\text{UC}}/v_{\text{c}} = \frac{3\pi^2}{16\phi_{\text{c}}}. \tag{6.25}$$

If there is no overlap of depletion zones (colloidal concentrations far from the close packing), the free volume for depletants is simply the volume unoccupied by the depletion zones. Overlap of the depletion zones lead to an increase of the free volume fraction for depletants. In the case of the platelets, overlap of the depletion zones occurs either from the side or from the flat phases of the hard disc. These two contributions can be conveniently split. Due to the different scalings of the unit cell, [26] overlap in the intracolumnar direction occurs at lower colloidal concentrations than in the intercolumnar one. One must account for the total number of overlaps: nine in the intercolumnar direction and three in the intracolumnar one. This allows to cast $\alpha_{\text{col}}^{\text{geo}}$ in a generic form:

$$\alpha_{\text{C}}^{\text{geo}} = \begin{cases} 1 - \frac{3v_{\text{excl}}^{\text{HP-PHS}}}{V_{\text{UC}}} & \text{if } \phi_{\text{c}} < \phi_{\text{c}}^{\parallel} \text{ (no overlap)}, \\ 1 - \left(\frac{3v_{\text{excl}}^{\text{HP-PHS}}}{V_{\text{UC}}} - \frac{3v_{\text{overl}}^{\parallel}}{V_{\text{UC}}}\right) & \text{if } \phi_{\text{c}}^{\parallel} \le \phi_{\text{c}} < \phi_{\text{c}}^{\perp} \text{ (overlap in } r_{\parallel}) \\ 1 - \left(\frac{3v_{\text{excl}}^{\text{HP-PHS}}}{V_{\text{UC}}} - \frac{3v_{\text{overl}}^{\parallel}}{V_{\text{UC}}} - \frac{9v_{\text{overl}}^{\perp}}{V_{\text{UC}}}\right) & \text{if } \phi_{\text{c}}^{\perp} \le \phi_{\text{c}} \text{ (overlap in } r_{\parallel} \text{ and } r_{\perp}), \end{cases} \tag{6.26}$$

where $\phi_{\text{c}}^{\parallel}$ is the solution of the equation

$$\widehat{\Delta}_{\parallel}(\phi_{\text{c}}) = 1 + q/\Lambda, \tag{6.27}$$

and $\phi_{\text{c}}^{\perp}$ is the solution of the equation

$$\widehat{\Delta}_{\perp}(\phi_{\text{c}}) = 1 + q. \tag{6.28}$$

Note first that the term $v_{\text{excl}}^{\text{HP-PHS}}$ accounts for the depletion zone volume [equivalent to Eq. (6.2) minus the term corresponding to the sphere volume, $2q^3/(3\Lambda^4)$]. Also note that with the approach described here we only account for two-body overlaps of the depletion zones, which is sufficient for small enough q-values. Further, in Eq. (6.26) the condition 'no free volume for depletants' is not shown for simplicity. Upon some algebra, these three different contributions read:

$$\begin{aligned}\frac{3v_{\text{excl}}^{\text{HP-PHS}}}{V_{\text{UC}}} &= \phi_{\text{c}}\left[(1+q)^2 + \frac{q}{6\Lambda}\left(6+3\pi q+4q^2\right)\right] \\ \frac{3v_{\text{overl}}^{\parallel}}{V_{\text{UC}}} &= \phi_{\text{c}}\left(1+\frac{q}{\Lambda}\right) - \frac{\pi}{2\sqrt{3}\widehat{\Delta}_{\perp}} + \frac{2\phi_{\text{c}}}{3}\left[3\Lambda A_2(q/(2\Lambda), \widehat{\Delta}_{\parallel}-1) + \Lambda^2\left(\widehat{\Delta}_{\parallel}-1\right)^3 + \frac{q^3}{\Lambda}\right] \\ \frac{9v_{\text{overl}}^{\perp}}{V_{\text{UC}}} &= \frac{\phi_{\text{c}}}{\pi}12A_2\left(\frac{1+q}{2}, \widehat{\Delta}_{\perp}\right),\end{aligned} \tag{6.29}$$

where A_2 is the area of intersect between two discs with radius R at a distance r:

$$A_2(R,r) = 2R^2\cos^{-1}\left(\frac{r}{2R}\right) - \frac{1}{2}r\sqrt{4R^2-r^2} \tag{6.30}$$

We shall finally note here that the algebraic complexity of Eq. (6.26) arises mostly due to the contribution to the depletion zone of the edges of the platelet.

6.C Direct Coexistence Monte Carlo Simulations

The (modified) direct Monte Carlo coexistence simulations consider hard colloidal platelets as hard oblate hard spherocylinders (OHSC) mixed with a non-adsorbing species which is modelled as penetrable hard spheres (PHSs). The direct colloid–colloid and depletant–colloid interactions are hard, whereas there are no depletant–depletant interactions. Simulations start with two non-equilibrated simulation boxes in contact. Each box contains either the C_1 or C_2 FVT-predicted coexistence volume fractions of colloids and polymers. The oblates are arranged in two columnar phases with the same column axes, whereas the depletants are distributed randomly without colloid–polymer overlaps. The whole simulation box contains N particles (discotics plus polymers), and a (MC) cycle is defined as N trials to displace and/or rotate a randomly chosen particle plus an attempt to change the aspect ratio of the simulation box (its volume is fixed). Two different equilibration steps are considered. Firstly, 1×10^6 cycles are conducted restricting the depletants to the volumes that they occupied in the initial configuration (equilibration of the discotic phases). Secondly, 3×10^6 cycles are carried out without restrictions (equilibration of the direct coexistence). Ensemble-averaged equilibrium colloid and polymer volume fractions are collected over the last 2×10^6 cycles. The method used here is a modified direct coexistence Monte Carlo [58] approach. It is applied, as far as we are aware of, for the first time to directly study isostructural coexistence in highly dense discotic systems explicitly accounting for a binary mixture (OHSCs and PHSs). These equilibrium direct-coexistence MC colloid and polymer volume fractions are used for two independent sets of simulations (1×10^6 cycles to equilibrate, plus 2×10^6 cycles for production), from which colloid–colloid and colloid–depletant distribution functions elucidate the structural details of the C_1 and C_2 phases. These direct coexistence simulations were blind-tested: two starting configurations at different colloid packing fractions in absence of depletants melt into a single one. Further details of the simulation method are beyond the scope of this thesis.

References

1. L. Yuan, X.L. Weng, J.L. Xie, L.J. Deng, Mater. Res. Innov. **19**, S1 (2015). https://doi.org/10.1179/1432891715Z.0000000001497
2. L. Bailey, H.N.W. Lekkerkerker, G.C. Maitland, Soft Matter **11**, 222 (2015). https://doi.org/10.1039/C4SM01717J
3. T. Ye, N. Phan-Thien, C.T. Lim, J. Biomech. **49**, 2255 (2016). https://doi.org/10.1016/j.jbiomech.2015.11.050
4. E. Dickinson, Food Hydrocoll. **52**, 497 (2016). https://doi.org/10.1016/j.foodhyd.2015.07.029
5. F.M. Van der Kooij, M. Vogel, H.N.W. Lekkerkerker, Phys. Rev. E **62**, 5397 (2000). http://journals.aps.org/pre/abstract/10.1103/PhysRevE.62.5397
6. W. Zhu, D. Sun, S. Liu, N. Wang, J. Zhang, L. Luan, Colloids Surf. A **301**, 106 (2007). http://www.sciencedirect.com/science/article/pii/S0927775706009666
7. L. Luan, W. Li, S. Liu, D. Sun, Langmuir **25**, 6349 (2009). https://doi.org/10.1021/la804023b
8. S.M. Oversteegen, C. Vonk, J.E.G.J. Wijnhoven, H.N.W. Lekkerkerker, Phys. Rev. E **71**, 041406 (2005). https://doi.org/10.1103/PhysRevE.71.041406
9. D. Kleshchanok, A.V. Petukhov, P. Holmqvist, D.V. Byelov, H.N.W. Lekkerkerker, Langmuir **26**, 13614 (2010). https://doi.org/10.1021/la101891e
10. D. Kleshchanok, J.-M. Meijer, A.V. Petukhov, G. Portale, H.N.W. Lekkerkerker, Soft Matter **7**, 2832 (2011). https://doi.org/10.1039/C0SM01206H
11. N. Doshi, G. Cinacchi, J.S. van Duijneveldt, T. Cosgrove, S.W. Prescott, I. Grillo, J. Phipps, D.I. Gittins, J. Phys.: Condens. Matter **23**, 194109 (2011). http://stacks.iop.org/0953-8984/23/i=19/a=194109
12. D. Kleshchanok, J.-M. Meijer, A.V. Petukhov, G. Portale, H.N.W. Lekkerkerker, Soft Matter **8**, 191 (2012). https://doi.org/10.1039/C1SM06535A
13. D. de las Heras, N. Doshi, T. Cosgrove, J. Phipps, D.I. Gittins, J.S. van Duijneveldt, M. Schmidt, Sci. Rep. **2**, 789 (2012). https://doi.org/10.1038/srep00789
14. J. Landman, E. Paineau, P. Davidson, I. Bihannic, L.J. Michot, A.-M. Philippse, A.V. Petukhov, H.N.W. Lekkerkerker, J. Phys. Chem. B **118**, 4913 (2014). https://doi.org/10.1021/jp500036v
15. M. Chen, H. Li, Y. Chen, A.F. Mejia, X. Wang, Z. Cheng, Soft Matter **11**, 5775 (2015). https://doi.org/10.1039/C5SM00615E
16. T. Nakato, Y. Yamashita, E. Mouri, K. Kuroda, Soft Matter **10**, 3161 (2014). https://doi.org/10.1039/C3SM52311J
17. M.A. Bates, D. Frenkel, Phys. Rev. E **62**, 5225 (2000). https://doi.org/10.1103/PhysRevE.62.5225
18. S.-D. Zhang, P.A. Reynolds, J.S. van Duijneveldt, J. Chem. Phys. **117**, 9947 (2002). https://doi.org/10.1063/1.1518007
19. S.-D. Zhang, P.A. Reynolds, J.S. van Duijneveldt, Mol. Phys. **100**, 3041 (2002). https://doi.org/10.1080/00268970210130146
20. L. Harnau, Mol. Phys. **106**, 1975 (2008). https://doi.org/10.1080/00268970802032301
21. D. de las Heras, M. Schmidt, Philos. Trans. R. Soc. A **371**, 20120259 (2013). https://doi.org/10.1098/rsta.2012.0259
22. R. Aliabadi, M. Moradi, S. Varga, J. Chem. Phys. **144**, 074902 (2016). https://doi.org/10.1063/1.4941981
23. A. Vrij, Pure Appl. Chem. **48**, 471 (1976). https://doi.org/10.1351/pac197648040471
24. R. Tuinier, G.J. Fleer, Macromolecules **37**, 8754 (2004). https://doi.org/10.1021/ma0485742
25. S.M. Oversteegen, R. Roth, J. Chem. Phys. **122**, 214502 (2005). https://doi.org/10.1063/1.1908765
26. H.H. Wensink, H.N.W. Lekkerkerker, Mol. Phys. **107**, 2111 (2009). https://doi.org/10.1080/00268970903160605
27. M.A. Bates, D. Frenkel, Phys. Rev. E **57**, 4824 (1998). https://journals.aps.org/pre/abstract/10.1103/PhysRevE.57.4824
28. M. Marechal, A. Cuetos, B. Martínez-Haya, M. Dijkstra, J. Chem. Phys. **134**, 094501 (2011). https://doi.org/10.1063/1.3552951

29. R. Tuinier, Adv. Condens. Matter Phys. (2016). https://www.hindawi.com/journals/acmp/2016/5871826/cta/
30. R. Tuinier, G.J. Fleer, J. Phys. Chem. B **110**, 20540 (2006). https://doi.org/10.1021/jp063650j
31. B. Widom, J. Phys. Chem. **77**, 2196 (1973). https://doi.org/10.1021/j100637a008
32. P.G. Bolhuis, M. Hagen, D. Frenkel, Phys. Rev. E **50**, 4880 (1994). https://journals.aps.org/pre/abstract/10.1103/PhysRevE.50.4880
33. M. Dijkstra, J.M. Brader, R. Evans, J. Phys.: Condens. Matter **11**, 10079 (1999)
34. K. Akahane, J. Russo, H. Tanaka, Nat. Commun. **7**, 12599 (2016). https://doi.org/10.1038/ncomms12599
35. V.F.D. Peters, M. Vis, Á. González García, R. Tuinier, in preparation (n.a.a)
36. M.A. Bates, D. Frenkel, J. Chem. Phys **110**, 6553 (1999). https://doi.org/10.1063/1.478558
37. H.H. Wensink, G.J. Vroege, J. Phys.: Condens. Matter **16**, S2015 (2004). http://stacks.iop.org/0953-8984/16/i=19/a=013
38. H.H. Wensink, H.N.W. Lekkerkerker, Europhys. Lett. **66**, 125 (2004). https://doi.org/10.1209/epl/i2003-10140-1
39. D. de las Heras, M. Schmidt, Soft Matter **9**, 8636 (2013). https://doi.org/10.1039/C3SM51491A
40. F.M. van der Kooij, K. Kassapidou, H.N.W. Lekkerkerker, Nature **406**, 868 (2000). https://doi.org/10.1038/35022535
41. S. Liu, J. Zhang, N. Wang, W. Liu, C. Zhang, D. Sun, Chem. Mater. **15**, 3240 (2003). https://doi.org/10.1021/cm034201o
42. V.F.D. Peters, M. Vis, H.H. Wensink, R. Tuinier, in preparation (n.a.b)
43. H.H. Wensink, Phys. Rev. Lett. **93**, 157801 (2004). https://doi.org/10.1103/PhysRevLett.93.157801
44. C.F. Tejero, A. Daanoun, H.N.W. Lekkerkerker, M. Baus, Phys. Rev. Lett. **73**, 752 (1994). https://journals.aps.org/prl/abstract/10.1103/PhysRevLett.73.752
45. G. Foffi, G.D. McCullagh, A. Lawlor, E. Zaccarelli, K.A. Dawson, F. Sciortino, P. Tartaglia, D. Pini, G. Stell, Phys. Rev. E **65**, 031407 (2002). https://doi.org/10.1103/PhysRevE.65.031407
46. C. Fernandes, A. Senos, Int. J. Refract. Met. Hard Mater. **29**, 405 (2011). http://www.sciencedirect.com/science/article/pii/S0263436811000333
47. D. Marenduzzo, K. Finan, P.R. Cook, J. Cell. Biol. **175**, 681 (2006). http://jcb.rupress.org/content/175/5/681
48. L. Sapir, D. Harries, Bunsen-Mag. **19**, 152 (2017). https://scholars.huji.ac.il/danielharries/publications/wisdom-crowd#
49. G. van Anders, D. Klotsa, N.K. Ahmed, M. Engel, S.C. Glotzer, Proc. Natl. Acad. Sci. USA **111**, E4812 (2014). https://doi.org/10.1073/Proc.Natl.Acad.Sci.U.S.A..1418159111
50. M. Dijkstra, in *Advances in Chemical Physics*, ed. by S.A. Rice, A.R. Dinner, vol. 156 (Wiley, Hoboken, 2014), Chap. 2. https://onlinelibrary.wiley.com/doi/10.1002/9781118949702.ch2
51. L. Onsager, Ann. N.Y. Acad. Sci. **51**, 627 (1949). https://doi.org/10.1111/j.1749-6632.1949.tb27296.x
52. H.H. Wensink, G.J. Vroege, H.N.W. Lekkerkerker, J. Phys. Chem B **105**, 10610 (2001). https://doi.org/10.1021/jp0105894
53. R. van Roij, Eur. J. Phys. **26**, S57 (2005). http://stacks.iop.org/0143-0807/26/i=5/a=S07
54. J.D. Parsons, Phys. Rev. A **19**, 1225 (1979). https://doi.org/10.1103/PhysRevA.19.1225
55. S.-D. Lee, J. Chem. Phys. **87**, 4972 (1987). https://doi.org/10.1063/1.452811
56. T. Odijk, Macromolecules **19**, 2313 (1986). https://doi.org/10.1021/ma00163a001
57. J.E. Lennard-Jones, A.F. Devonshire, Proc. R. Soc. A **163**, 53 (1937). https://www.jstor.org/stable/97067?seq=1#page_scan_tab_contents
58. A.Z. Panagiotopoulos, Mol. Phys. **61**, 813 (1987). https://doi.org/10.1080/00268978700101491

Part III
Spherical Association Colloids

Chapter 7
On the Colloidal Stability of Association Colloids

7.1 Introduction

The association of polymers and surfactants (macromolecules) into soft colloidal particles provides a playground for generating a wide range of self-assembled architectures in selective solvents [1]. Substantial attention has been paid to predetermine the preferred morphology of association colloids [2]. Among the possible micellar shapes, the spherical one is appealing due to its wide applicability for instance in coatings [3], in food [4], and as drug delivery systems [5–8]. In many applications, control is not only desired over the morphology of the self-assembled structure, but also over the thermodynamic stability of the micellar suspension. A widely applied technique to enhance the stability of *inorganic* colloidal particles is grafting polymers onto their surface, which leads to steric stabilisation [9, 10]. For spherical micelles formed by block copolymers, such steric stabilisation is inherent [11]. Understanding how micelles interact is key to envisage, and therefore predict, the stability of a micellar solution. Previously presented models for micelle–micelle interactions account for the core as a hard surface onto which the solvophilic components are tethered [12–14]. Micelles are however responsive, since the assembled molecules are, presumably, in equilibrium with free ones, so the core–corona interface is soft and adaptable [15–17]. For micelles with large coronal domains, the interaction between micelles mediated by overlap of coronas has been compared to that of starlike polymers [18, 19]. It is noted, however, that starlike polymers are not self-assembled structures [20], and such models hence neglect the presence of free diblock copolymer in solution. For this reason, the interaction between diblock copolymer spherical micelles may be quantified while accounting for their soft polymeric and associative nature. When computing the micelle–micelle interaction potential, we allow the aggregation number (g_p, the number of polymers composing the micelle) to equilibrate with free polymer in the bulk at each intermicelle separation distance r. We account for intermicellar distances $r \geq 2R_\mathrm{h}^\mathrm{o}$, with R_h^o the hydrodynamic radius of an undistorted micelle. Hence, we focus on dilute micelle suspensions rather than on

© Springer Nature Switzerland AG 2019

Á. González García, *Polymer-Mediated Phase Stability of Colloids*,
Springer Theses, https://doi.org/10.1007/978-3-030-33683-7_7

Table 7.1 System parameters chosen in this study: number of solvophobic (m) and solvophilic (n) repeating units, and Flory–Huggins interaction parameters. A and B refer to a solvophilic and solvophobic segment, respectively; W is a solvent molecule

m	n	χ_{BW}	χ_{AW}	χ_{AB}
16, 24, 32	varied	2, 3	varied	1

high-density solid phases of micelles [18] or possible micelle morphology transformations above their overlap concentration [21, 22].

A useful indicator for the thermodynamic stability of a colloidal suspension is the second osmotic virial coefficient B_2 [23, 24]. Yet, experimentally measured B_2 values for micelle suspensions are limited [25–29]. The value of B_2 can be used to specify the (colloidal) stability of a suspension. For a collection of hard spheres, $B_2^* \equiv B_2/v_c = 4$ [30], where v_c is the volume of the colloidal particle considered. If repulsive forces beyond the pure hard core excluded volume interaction are present between the colloidal particles, $B_2^* > 4$. For monocomponent systems of interacting spheres, the Vliegenthart–Lekkerkerker (VL) criterion [31] identifies the onset of colloidal gas–liquid coexistence at $B_2^* \lesssim -6$ [32, 33]. Both the hard sphere limit and the VL criterion are used here as *indicative* values for the colloidal stability (see Sect. 1.3.2).

In this chapter, we present a bottom-up approach to study the stability of a micellar suspension of diblock copolymers. We estimate micelle–micelle interactions via self-consistent field calculations for block copolymers with different block lengths and solubilities. Results are compared with an alternative analytical expression for the interaction potential based upon the thermodynamics of micelle formation [34]. Here, we investigate whether one can describe self-consistent field (SCF) results with a theoretical model. Furthermore, we calculate the normalised second virial coefficient, B_2/v_c, considering v_c as the volume of an isolated micelle, and evaluate its dependency on block copolymer composition and solvency parameters of the blocks. The SCF-approach followed was explained in Chap. 1, hence we focus here on the results. For completeness, the set of closed expressions used to compare the SCF results is presented in Sect. 7.2. The set of system parameters used is summarised in Table 7.1.

7.2 Semi-analytical Expression for the Interaction Potential

An analytical expression for micelle–micelle interactions can be obtained from previously developed theories for block copolymer micelles. The presence of K surrounding micelles exerts an isotropic compression on a central micelle at small enough inter-micelle distances r. The confined (central) micelle is assumed to be in equilibrium with free copolymers. This enables to minimize the unfavourable increase of the free energy upon compression (e.g., the aggregation number g_p is allowed to vary with r). The micelle–micelle interaction $W(r)$ can be expressed as:

$$W(r) = \frac{2}{K}\left[f_{\text{mic}}(r) - f_{\text{mic}}(r = \infty)\right], \tag{7.1}$$

where $f_{\text{mic}}(r)$ is the free energy of a micelle whose centre is separated a distance r from a neighbouring one. Hence, $f_{\text{mic}}(r = \infty)$ is the free energy of an isolated micelle. The free energy of a micelle can be approximated as the sum of three contributions: the elastic free energy of the core-forming blocks, the elastic free energy of the corona-forming blocks and the interfacial energy between the core and the solvent at the core–corona interface. We use an approximate expression for $f_{\text{mic}}(r) = F_{\text{mic}}(r)/(k_B T)$ (with k_B being Boltzmann's constant and T the absolute temperature) by modifying a result from Zhulina and Borisov [34]:

$$f_{\text{mic}}(r) = \underbrace{g_{\text{p}}(r)\frac{3\pi^2 R_{\text{c}}^2(r)}{80b^2 m}}_{\text{elastic, core}} + \underbrace{\frac{g_{\text{p}}(r)^{3/2}}{2\sqrt{\pi}}\ln\left[1 + \frac{T(r)}{R_{\text{c}}(r)}\right]}_{\text{elastic, corona}} + \underbrace{\frac{4\pi\widetilde{\gamma}}{b^2}\left[R_{\text{c}}(r)^2 - g_{\text{p}}(r)R_{\text{B}}^2\right]}_{\text{core-corona interface}}. \tag{7.2}$$

Here b is the size of a monomer, $R_{\text{c}}(r)$ and $T(r)$ represent the core radius and the corona thickness respectively, R_{B} is the radius of the collapsed B block in an unassembled block copolymer molecule and $\widetilde{\gamma}$ is the (normalised) interfacial tension between the core and the solvent. The value of $\widetilde{\gamma}$ is approximated from the Helfand–Tagami equation [35]:

$$\widetilde{\gamma} = \sqrt{\chi_{\text{BW}}/6}. \tag{7.3}$$

Both $R_{\text{c}}(r)$ and $T(r)$ can be expressed as a function of $g_{\text{p}}(r)$. In a spherical micelle composed of g_{p} copolymer molecules the volume of the spherical hydrophobic core is given by

$$V_{\text{C}} = g_{\text{p}} V_{\text{B}} = \frac{g_{\text{p}} m b^3}{\phi_{\text{B}}}, \tag{7.4}$$

where V_{B} is the volume occupied by a single B block and ϕ_{B} is the polymer volume fraction in the core. For values of $\chi_{\text{BW}} > 1$, ϕ_{B} is close to unity and can be approximated as:

$$\phi_{\text{B}} = 1 - e^{-\frac{4}{3}\chi_{\text{BW}}}. \tag{7.5}$$

The expression above matches reasonably well with SCF computations [36]. Hence, R_{c} and R_{B} can be written as:

$$\frac{R_{\text{B}}}{b} = \left(\frac{3m}{4\pi\phi_{\text{B}}}\right)^{1/3}, \qquad \frac{R_{\text{c}}}{b} = \left(\frac{3g_{\text{p}}m}{4\pi\phi_{\text{B}}}\right)^{1/3}. \tag{7.6}$$

To estimate the total radius of the micelle R we assume that the volume of each A block does not change upon assembly, and equals the unperturbed volume V_A, given by

$$V_A = \frac{4\pi}{3} R_{g,A}^3, \tag{7.7}$$

where $R_{g,A}$ is the gyration radius of the A block, which is estimated from [37]:

$$\frac{R_{g,A}}{b} = 0.31 n^{1/2} \left[1 + \sqrt{1 + 6.5(1 - 2\chi_{AW}) n^{1/2}} \right]^{0.352}. \tag{7.8}$$

The total radius of the micelle and the corona thickness are thus given by

$$R = \left(\frac{3 g_p (V_A + V_B)}{4\pi} \right)^{1/3}, \qquad T = R - R_c. \tag{7.9}$$

Hence, given the χ-parameters and the block copolymer composition we only need $g_p(r)$ to calculate $W(r)$. The results of this semi-analytical theory are denoted as hybrid because SCF input is needed [only through $g_p(r)$].

7.3 Results and Discussion

First, we evaluate the dependence of the equilibrium micelle properties on the intermicelle distance. This yields micelle–micelle interactions, for which case examples are presented. Subsequently, we use these micelle–micelle interaction potential to compute the second virial coefficients mediated by the solvophilic block solubility and chain length. Finally, the effect of the diblock copolymer composition and monomeric interaction parameters on the colloidal phase stability is summarised into two comprehensible plots.

7.3.1 Equilibrium Properties of Micelles with Varying Intermicelle Distance

We focus first on the changes of the micellar equilibrium properties at different intermicelle separation distances r. These micellar properties were studied using a lattice with concentration gradients in one (spherical lattice) or two directions (cylindrical lattice). In Fig. 7.1 we present the grand-potential Ω obtained via Scheutjens–Fleer self-consistent, mean field (SCF) computations [38–40] as a function of the aggregation number g_p for different lattices for micelles formed by diblock copolymers $B_{24}A_{45}$ in a solvent W. The interaction between blocks and of blocks with the

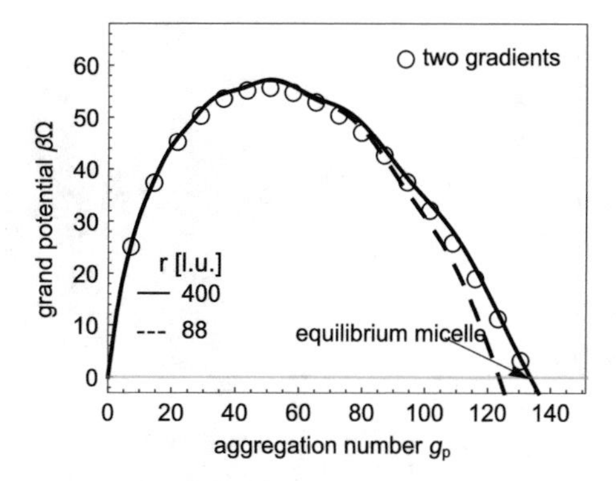

Fig. 7.1 Grand potential from SCF computations Ω as a function of the aggregation number g_p at two different intermicelle distances r obtained using a spherical lattice (solid and dashed curves). Open circles correspond to the grand-potential curve considering a single micelle in a cylindrical lattice, allowing to study spherical micelles with concentration gradients in two directions. The micelle considered is composed of $B_{24}A_{45}$ diblock copolymers with $\chi_{BW} = 2$, $\chi_{AW} = 0.4$, and $\chi_{AB} = 1$

solvent are specified via Flory–Huggins interaction parameters, namely $\chi_{BW} = 2$ and $\chi_{AW} = 0.4$; $\chi_{AB} = 1$ is used in all our calculations. The grand-potential curves using one or two concentration gradients practically overlap if the lattice dimensions are large enough; in such a case the dilute solution limit of individual micelles is reached. There are no appreciable differences in the maximum free energy required to form a micelle as micelles get closer to each other (decreasing r). However, the average equilibrium aggregation number of the micelle (which satisfies $\Omega = 0$ with $\partial\Omega/\partial g_p < 0$) decreases when micelles are formed at small enough distances. For all diblock sequences B_mA_n studied here, it was verified that the preferred self-assembled structure is a spherical micelle.

The corresponding equilibrium concentration profiles are presented Fig. 7.2. If the number of lattice sites (N_{lat}) is sufficiently large, the micelle size and aggregation numbers are independent of the lattice type considered (spherical or cylindrical lattice). This can be appreciated by the projection of the equilibrium sizes from the spherical lattice onto the cylindrical one (left panel of the top row in Fig. 7.2). All sizes are expressed in terms of lattice units [l.u.]. Note that the distance between the centres of the micelles r is set by the number of lattice sites. For one-gradient SCF computations, $r = 2N_{lat}$ (with N_{lat} the number of concentration shells considered). For two-gradient computations, $r = 2N_{lat}^{r}$ (if the nearest micelles are in the radial direction) or $r = N_{lat}^{y}$ (in case the nearest micelles are in the longitudinal direction). From the concentration profiles, the hydrodynamic micelle radius (R_h) can be computed (see Chap. 1). In the dilute limit, $r \gg 2R_h^o$, $R_h^o \approx 20$ [l.u.] for the $B_{24}A_{45}$ block copolymer micelle. The superscript 'o' denotes micelle undistorted properties. The solvophobic blocks are concentrated in the core of the spherical micelle, which is compact and nearly solvent-free [41]. The approximated core size R_c is indicated via dashed vertical lines. The solvophilic blocks are mainly located in the corona,

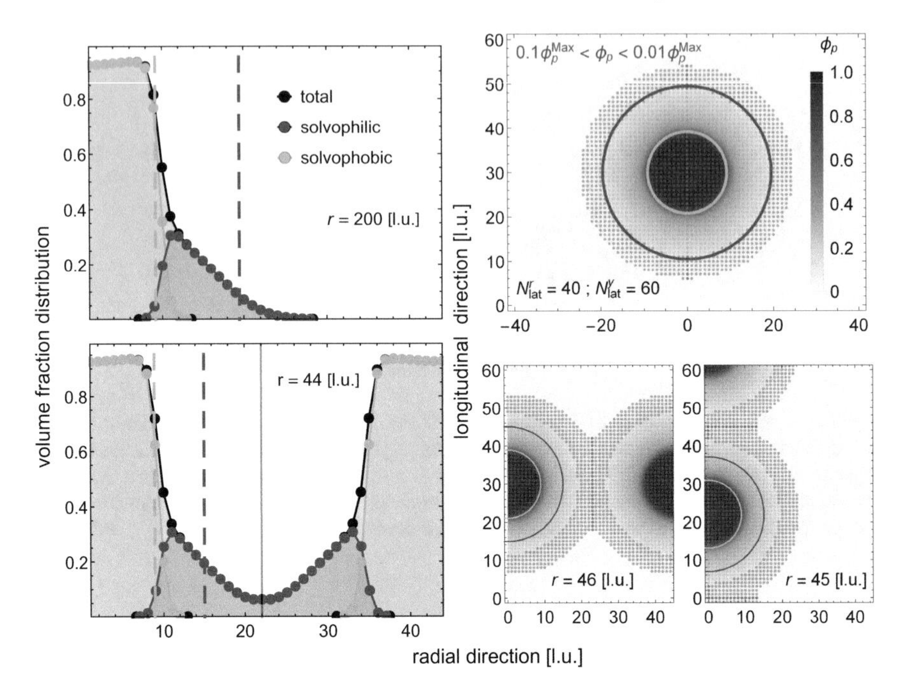

Fig. 7.2 Concentration profiles computed using SCF theory in a spherical lattice with concentration in one dimension (left panels) or in a cylindrical lattice with concentration gradients in two dimensions (right panels) at equilibrium conditions (see Fig. 7.1). Nearest-neighbour micelle distances r are indicated, as well as the number of lattice sites (N_{lat}). Vertical lines correspond either to the corresponding hydrodynamic sizes (dashed) or to the mirror (solid, only in the bottom leftmost panel). Dark and light grey circles (right panels) correspond to the sizes on the left, while clouds of points correspond to polymer concentrations in between 1% and 10%

which is well-solvated. It is noted that solvophilic polymer segments are also significantly present at positions beyond R_{h}^{o}, see top panels of Fig. 7.2. We denote the region where solvophilic segments are clearly present ($\phi_{\text{p}} \gg \phi_{\text{p}}^{\text{bulk}}$, with $\phi_{\text{p}}^{\text{bulk}}$ the polymer bulk concentration) beyond R_{h}^{o} as the *solvophilic tails*. When micelles get close, the overlapping of these outer tail regions leads to a contraction of the coronas already at intermicelle distances $r > 2R_{\text{h}}^{\text{o}}$. This induces a decrease in the micelle size with respect to the dilute limit at $r > 2R_{\text{h}}^{\text{o}}$ (see Fig. 7.3). The interpenetration of these solvophilic tails is clearly visible in the bottom right panels of Fig. 7.2, where the density profiles from the two concentration gradient computations in either the radial or the longitudinal length is of the order of $2R_{\text{h}}^{\text{o}}$.

We use the micelle hydrodynamic size in the dilute limit as characteristic length scale for the pair interaction between micelles. This quantity relates with theoretical predictions for diblock copolymer micelles [42]. The variation of the aggregation number as a function of the normalised intermicelle distance $\tilde{r} = r/2R_{\text{h}}^{\text{o}}$ is shown in Fig. 7.3. Due to the lattice-nature of the approach followed, it is useful to compare the aggregation number change (Δg_{p}) normalised by the number of nearest neighbouring

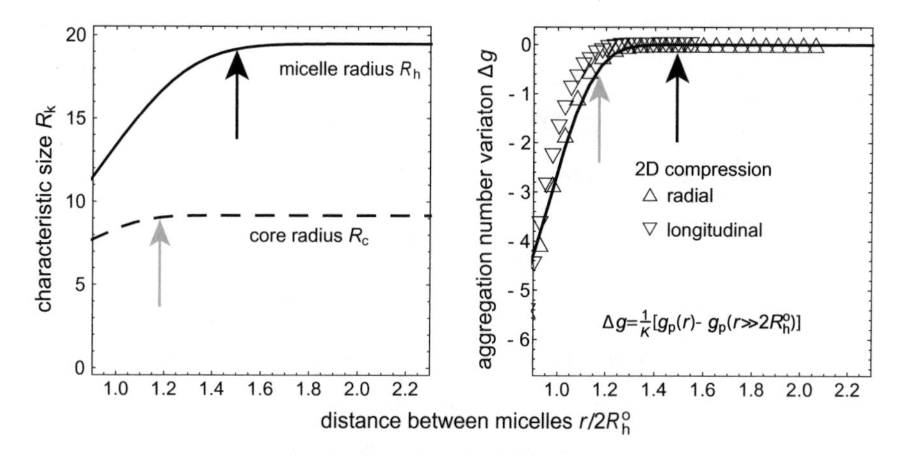

Fig. 7.3 Characteristic sizes (left panel) and effective change of aggregation number (right panel) as a function of the normalised intermicelle distance $r/2R_h^o$ for the same system parameters as in Fig. 7.1. Arrows are used for the estimated onsets of corona (black) and core (grey) compressions

micelles K when bringing the micelles closer to each other:

$$\Delta g_p = \frac{1}{K}\left[g_p(r) - g_p(r \gg 2R_h^o)\right]. \tag{7.10}$$

As micelles get closer (decreasing r) their sizes decrease due to the overlap of the solvophilic tails, which lead to contraction of the coronas. From the results in Fig. 7.3 it follows that both the characteristic size of the micelle core and corona as well as the aggregation number decrease with decreasing r. The overall size decreases already for $r \lesssim 3.2R_h^o$: the outer solvophilic tails start to interact significantly near $\tilde{r} \approx 1.6$. The onset of the decrease of the core size and aggregation number appear simultaneously near $\tilde{r} \approx 1.2$. This may be explained by the strong dependence of the aggregation number on the core and corona-forming block size [34]: for $\tilde{r} \lesssim 1.2$ the core is compressed, and diblock copolymers start to dissociate from the micelle.

The decrease of g_p as micelles get closer contrasts with regular scaling models, where g_p is assumed to remain constant up to the limit where micelles overlap, and increases beyond overlap of the micelles [34]. We note that the model presented here concerns *dilute* suspensions of micelles, as we do not study the micellar changes for $r < 2R_h^o$. Upon approaching overlap of micelles, a small decrease of the aggregation number (of the order of what we find here) might be inferred from experiments [43]. The trend shown in Fig. 7.3 holds when increasing the solvophilic block length, but the variation in g_p is smaller with increasing n at fixed $\tilde{r}$ (as micelles become more starlike). It has been suggested that the increase of g_p above overlap of the micelles is associated with a change of the preferred micellar morphology far beyond micelle overlap concentration [21, 22]. The 2D-gradient SCF approach followed here might be able to capture such effects, which are out of the scope of the present study.

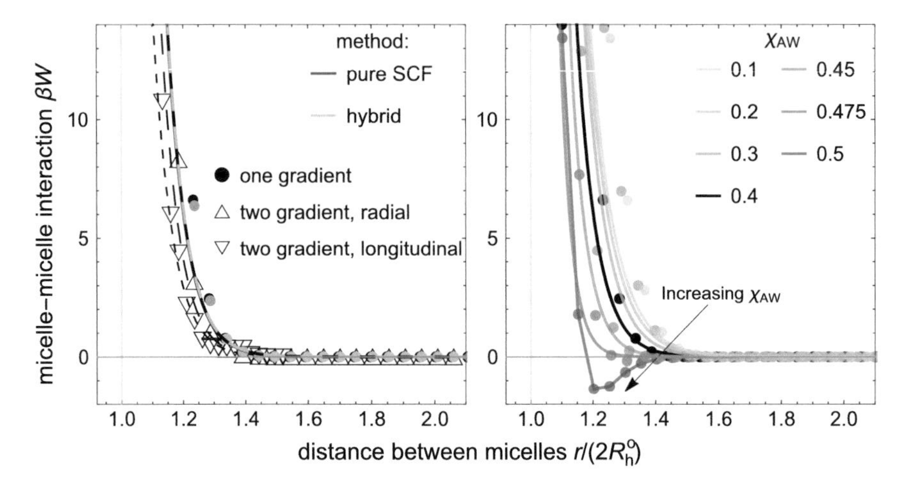

Fig. 7.4 Left panel: interaction potential considering different pure SCF approaches and the hybrid model [Eq. (7.1)], where the values of g_p are obtained from SCF. Curves correspond to the hard-core-Yukawa (HCY) potential fit while symbols are calculated points. Solid, black curve correspond to the HCY fit of the one concentration gradient calculations. Dashed grey curve corresponds to the hybrid method, overlaps with the solid black curve. Dashed black curves hold for the two-gradient computations where micelles are brought closer to each other in the radial or longitudinal direction. Right panel: interaction potential between micelles for various solvent quality parameters of the corona-forming block. Diblock copolymer considered is: $B_{24}A_{45}$, $\chi_{BW} = 2$, $\chi_{AB} = 1$, and varying χ_{AW} as indicated. Curves correspond to the hard-core-Yukawa potential fit while symbols are extracted from SCF data. For $\chi_{AW} = 0.5$, the calculated points are simply joined as a HCY fit is not applicable in this case

7.3.2 Model Comparison and Lattice Geometry Effects

In this section, micelle–micelle interactions obtained via the different approaches, and using different lattice types are compared. We consider first diblock copolymers $B_{24}A_{45}$ with parameters $\chi_{BW} = 2$, $\chi_{AW} = 0.4$, $\chi_{AB} = 1$. The dependence of the micelle equilibrium with intermicelle distance provide all required components for calculating these micelle–micelle interactions (see Sect. 1.3.5), which are presented in Fig. 7.4. We first consider purely SCF lattice computations using one or two concentration gradients, as well as a semi-analytical approach (see Sect. 7.2) in which the only input from the SCF computations is the change in $g_p(r)$ (hybrid). The different methods produce very similar results: a strong, short-range repulsion takes place at short intermicelle distances ($\tilde{r} \lesssim 1.5$) which originates from a significant excluded volume repulsion of the solvophilic tails, corresponding to the situation where coronas contract (see left panel of Fig. 7.3). This repulsive interaction is similar to a brush-like repulsion between polymer-grafted colloids [12, 13, 44] and starlike polymers [45]. For spherical micelles, however, we find that the 'surface' at which the 'brushes' (solvophilic tails) are grafted is soft, and the effective grafting density is adaptative: both g_p and R_h depend on r. The SCF computations account for the soft and adaptable nature of the micelles.

Table 7.2 HYC-fitted range of interaction (q_Y), extrapolated contact potential [$W(r) = 2R_h^o$], and normalised second virial coefficient (B_2^*) for a collection of micelle–micelle interactions obtained from the SCF approach and using Eq. (7.1), where the values of g_p are calculated via SCF-computations (hybrid). Diblock considered is $B_{24}A_{45}$; the FH interaction parameters are $\chi_{BW} = 2$, $\chi_{AW} = 0.4$ and $\chi_{AB} = 1$

Method	K	q_Y	$W(r = 2R_h^o)$	B_2^*
One-gradient SCF	12	0.13	90.1	9.9
Equation (7.1) + SCF (hybrid)	12	0.13	90.2	9.9
Two-gradients SCF, radial compression	6	0.12	85.5	9.2
Two-gradients SCF, longitudinal compression	2	0.13	62.7	8.9

The micelle–micelle interactions obtained via the pure SCF (either one or two concentration gradients) and the hybrid method are quite close (left panel of Fig. 7.4). It appeared to be convenient to fit the interaction potential via a hard-core Yukawa (HCY) interaction (see Sect. 1.3.1). This allows to systematically quantify the range of repulsion (q_Y) between micelles and how the interaction depends on the diblock copolymer properties. Further, the HCY model has been proposed as a model potential for the interaction between block copolymer micelles [46]. The fitting results are presented in Table 7.2. The fitted HCY curves can describe the SCF data points quite well, see Fig. 7.4. Variations are expected in the contact potential values for the different approaches due to the steepness of the interactions calculated. The q_Y-values obtained do however not vary significantly. The small differences can be related to slightly different $\Delta g_p(r)$ values, see Fig. 7.3. Our bottom-up approach differs from the ones previously reported in literature, where the mapping of the micelle–micelle repulsion into a HCY [46] or pure hard sphere [47] pair potential was performed using a top–down approach, via fitting experimentally collected structure factors with theoretical ones.

To gain more insight into the colloidal phase stability of micelle suspensions we also compare the obtained normalised second virial coefficient $B_2^* = B_2/v_c$, where the effective colloidal particle volume v_c is taken as the hydrodynamic volume of a micelle in the dilute solution limit:

$$v_c = \frac{4\pi}{3}\left(R_h^o\right)^3. \tag{7.11}$$

Details on the calculation of B_2^* from the pair interaction can be found in Chap. 1. The slight decrease of B_2^* with decreasing the number of nearest neighbour micelles K points towards the small overestimation of the contact potential values when calculations on the spherical lattice are conducted, most likely due to the different core compressions induced. Deviations of the results depending on the number of concentration gradients using SCF computations are expected [13, 48]. However, SCF calculations in the spherical lattice are sufficiently accurate to resolve the main characteristics of diblock copolymer micelle-micelle interactions.

Table 7.3 HYC-fitted range of interaction (q_Y), extrapolated contact potential [$W(r) = 2R_h^o$], and normalised second virial coefficient (B_2^*) for a collection of micelle–micelle interactions obtained from the SCF approach and using Eq. (7.1), where the values of g_p are calculated via SCF-computations (hybrid). Diblock considered is $B_{24}A_{45}$; the FH interaction parameters are $\chi_{BW} = 2$, $\chi_{AB} = 1$, and χ_{AW} is varied as indicated

χ_{AW}	SCF			Hybrid		
	q_Y^{SCF}	$W(r = 2R_h^o)$	B_2^*	q_Y^{hyb}	$W(r = 2R_h^o)$	B_2^*
0.1	0.17	76.2	12.3	0.17	83.4	12.4
0.2	0.16	82.4	11.8	0.16	86.8	11.9
0.3	0.14	90.6	11.0	0.14	86.1	11.0
0.4	0.13	91.5	9.9	0.13	88.2	9.9
0.45	0.11	90.1	8.6	0.11	86.1	8.6
0.475	0.08	87.3	7.1	0.08	82.2	7.1
0.5	n.a.	n.a.	1.0	n.a.	n.a.	1.6

7.3.3 *Influence of Coronal Block Solvency*

Next, we discuss the effects of the solvophilic block solvency parameter (χ_{AW}) on the inter-micellar interactions of the same block-copolymer type as before ($B_{24}A_{45}$ with $\chi_{BW} = 2$, and varying χ_{AW}). This solvency parameter governs the colloidal phase stability of the micellar suspension (as shown in the next section). We consider SCF computations with concentration gradients in one direction. In Fig. 7.4 (right panel) micelle–micelle interactions with varying χ_{AW}-values are shown. The black curve corresponds to $\chi_{AW} = 0.4$, the reference situation reported already in the previous section. For $\chi_{AW} < 0.4$, the repulsions get more long-ranged which increases B_2^* (see also Table 7.3). By increasing the solvent quality for the corona-forming blocks, the tails extend further from R_h^o. This leads to a longer-ranged repulsion. The opposite trend is observed for $\chi_{AW} > 0.4$. Strikingly, a shallow attraction between the micelles around $\tilde{r} \approx 1.2$ appears for $\chi_{AW} = 0.5$. At the Θ-solvent conditions, the excluded volume between corona-forming segments is compensated by the attraction among them. When the corona blocks start to overlap, these attractions become increasingly important as there are less corona-solvent contacts. This explains, we think, the attractive part of the potential for $\chi_{AW} = 0.5$. The repulsion contribution at Θ-solvent conditions arises from compression of the core. Upon further increase of χ_{AW} the attractive part of the potential would increase.

In Table 7.3 we present the obtained range of repulsion (when possible) as well as the normalised second virial coefficient for the potentials in Fig. 7.4 (right panel). With increasing χ_{AW} the q_Y values get smaller, and B_2^* decreases. Near $\chi_{AW} \approx 0.5$ the colloidal phase stability of the micelle suspension drops strongly. The pair interactions (hence their fitting parameters) do not significantly vary with the method (hybrid method or pure SCF) used in their calculation.

Table 7.4 Similar to Table 7.3 but for various chain lengths n for the diblock $B_{24}A_n$; the FH interaction parameters are $\chi_{BW} = 2$, $\chi_{AB} = 1$, and $\chi_{AW} = 0.4$

n	SCF			Hybrid		
	q_Y	$W(r = 2R_h^o)$	B_2^*	q_Y	$W(r = 2R_h^o)$	B_2^*
45	0.128	91.6	9.8	0.128	87.5	9.8
90	0.148	48.5	9.8	0.148	48.6	9.8
180	0.165	25.1	9.4	0.164	26.7	9.5
225	0.167	21.8	9.2	0.166	23.8	9.4
450	0.180	11.1	8.4	0.178	13.1	8.7

7.3.4 Solvophilic Block Length Effects

In this section we address the effect of varying the solvophilic block length, which leads to an increase of the coronal thickness. The effect of increasing the the thickness of this peripheral region is two-fold. On the one hand, due to a soft decay of the solvophilic tails, the steric repulsion gets more long-ranged (concentration profiles on the left panel of Fig. 7.5). This leads to a larger q_Y-value, see Table 7.4. On the other hand, the aggregation number decreases with increasing R_h^o due to an increased overall diblock solvency: a micelles coexist with a higher diblock bulk concentration. Thus, the effective grafting density of solvophilic tails from the core decreases with increasing n. As observed in the previous section, the strongest contribution to the steric repulsion between micelles arises, within our model, from compression of the core. Therefore, there is a balance between the range and the strength of the steric repulsion due to the coronal decays with increasing the solvophilic block length n, which leads to an overall high B_2^* value which weakly depends on the particular n-value. When increasing n, the decrease of g_p weakens: the core compression gets more screened upon increasing the corona thickness.

In the right panel of Fig. 7.5, examples of micelle–micelle interactions for various n-values ($B_{24}A_n$ with $\chi_{BW} = 2$, $\chi_{AW} = 0.4$, and $\chi_{AB} = 1$; black curve corresponds to the chosen reference diblock) are plotted. The shape of the interaction potentials resemble those presented in Fig. 7.4 (left panel). As can be appreciated, the HCY potential fits even better for large n-values: the interaction between diblock copolymeric micelles (particularly, with long hydrophilic tails) is similar to the outer-soft-core contribution of the interaction between starlike polymers [18, 20]. In Table 7.4 the resulting HCY-fitted interaction range, contact potential, and normalised second virial coefficient are listed for various hydrophilic block lengths n. For the conditions investigated B_2^* still remains approximately constant (though a slight decrease is appreciated when considering very large coronal domains). Note that q_Y is defined relative to the contact potential (see Sect. 1.3.1).

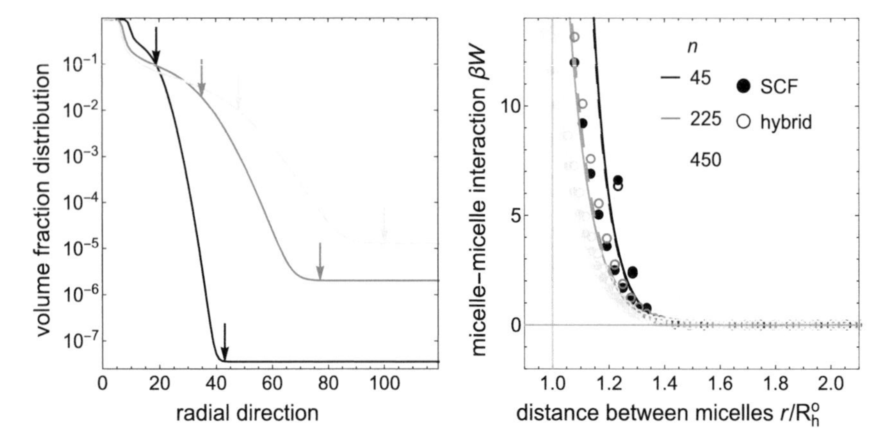

Fig. 7.5 Left panel: polymer segment concentration profiles computed using SCF theory in a spherical lattice with concentration gradients in one direction for diblocks of type $B_{24}A_n$ (with $\chi_{BW} = 2$, $\chi_{AW} = 0.4$, and $\chi_{AB} = 1$). Arrows in the top-left quadrant correspond to the hydrodynamic radius of the micelle, whereas arrows at low concentrations denote the position at which concentration has decay to 1.005 times its bulk value. Right panel: interaction potential between micelles for the same diblocks as on the left panel. Solid curves correspond to the hard-core-Yukawa potential fit of the SCF data, whilst dashed curves correspond to fittings of the hybrid approach presented

7.3.5 *Colloidal Phase Stability of Spherical Micelles*

In this section a colloidal phase stability overview of diblock copolymer micelle suspensions is presented in terms of the calculated second virial coefficient. This quantity can be related to the colloidal phase stability and can be experimentally measured using light-scattering techniques. The interaction between polymer brushes anchored to solid surfaces (steric stabilisation) sensitively depends on the grafting density of polymers [12, 49, 50]. To compare the interactions between colloidal spheres with anchored polymeric brushes and those between spherical copolymer micelles, we considered an effective grafting density of solvophilic blocks at the core-corona interface:

$$\Gamma_c = \frac{g_p}{4\pi R_c^2}, \tag{7.12}$$

where R_c is the core size, estimated from the SCF concentration profiles. The absolute value of Γ_c for diblock copolymer micelles depends on the considered system parameters: the number of solvophobic block segments (m), the number of solvophilic block segments (n), and their solvency parameters (χ_{BW} and χ_{AW}). However, in terms of the colloidal phase stability, the solvency and length of the core-forming blocks (m and χ_{BW}) does hardly affect B_2^* (as shown in the left panel of Fig. 7.6) for the spherical micelles studied. This is due to a balance between the range (increasing with n) and the strength (decreasing with n) of the steric repulsion. In fact, a value of

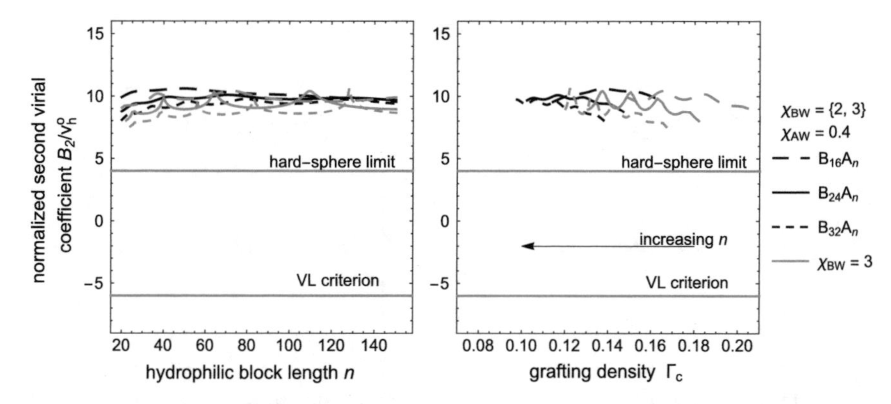

Fig. 7.6 Left panel: normalised second virial coefficient B_2^* of spherical micelles composed of diblock copolymers $B_{24}A_n$ with increasing solvophilic block length n. Right panel: B_2^* for the same data sets as on the left panel but in terms of the grafting density of solvophilic segments from the core Γ_c

$B_2^* \approx 9 \pm 1$ is found independently of $\{m, n, \chi_{BW}\}$ for a fixed coronal block solvency of $\chi_{AW} = 0.4$ (see Fig. 7.6). Hence, the grafting density at the core–corona interface (set by the diblock properties) hardly mediates repulsive micelle–micelle interactions upon variations of the solvophobic block (within the range where spherical micelles are preferred over other micelle morphologies). This is in contrast with the expectations for the interaction between hard spheres with anchored brushes. Next, we focus again on the influence of the interaction between the solvophilic tails as mediated by χ_{AW}.

The influence of the corona block solvency (χ_{AW}) on B_2^* is plotted in Fig. 7.7. For $\chi_{AW} \leq 0.45$, we find $B_2^* > 4$ for all values of n, indicating that micelles (with these characteristics) always interact in an overall repulsive fashion. The value of χ_{AW} does not only affect the grafting density, but also the interaction between the coronal tails. Upon approaching Θ-solvent conditions ($\chi_{AW} \to 0.5$) the *mutual* excluded volume repulsion decreases. Thus, contrary to what is expected from sterically stabilized *inorganic* colloids [12, 49], B_2^* *decreases* due to solvophobic effects with increasing the (diblock properties dependent) grafting density.

The attractive part of the pair interaction may be deep enough to destabilize the micelle suspension (Fig. 7.4, right panel), as seen in Fig. 7.7 (left panel) for $\chi_{AW} \gtrsim 0.5$. In contrast with theoretical predictions for polymer-grafted colloids [12, 49], this colloidal destabilisation arises (within our model) without considering direct attractions between micelles. Colloidal destabilisation around Θ-solvent conditions for the corona arises due to solvophobic effects: the enthalpic gain due to the solvent expel as micelles get closer is sufficient to compensate the entropic penalty of compressing the solvophilic tails.

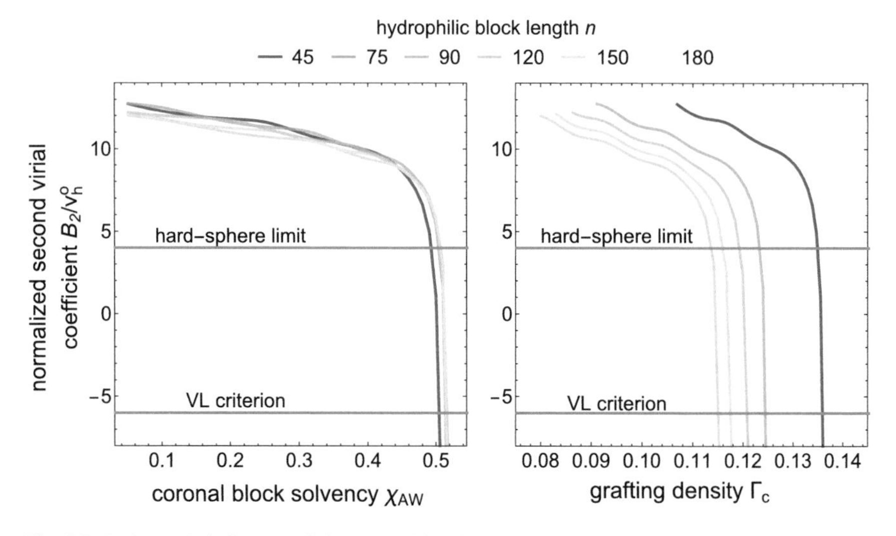

Fig. 7.7 Left panel: influence of the coronal block solvency (via χ_{AW}) on the normalised second virial coefficient B_2^* of spherical micelles composed of diblock copolymers $B_{24}A_n$ for a collection of different solvophilic block lengths n. Right panel: B_2^* for the same data sets as on the left but in terms of the grafting density of solvophilic segments at the core-corona interface

7.4 Conclusions

In this chapter the interaction between dilute (diblock) copolymer micelles is quantified using numerical self-consistent field (SCF) computations and analytical theory. We use the aggregation number obtained from the SCF computations as an input for the (semi-)analytical theory. The micelle–micelle interactions obtained via the two methods are in good agreement, also when considering different lattice topologies. Particularly, the range of the interaction and the normalised second virial coefficient are all rather similar: they are not sensitive to the method used and to the amount of concentration gradients that are considered. In our approach, we account for the soft and associative nature of these diblock copolymer micelles as they get closer. At each condition the equilibrium micellisation is re-evaluated: all polymer blocks in the micelles remain associative and fully responsive and can conformationally rearrange and equilibrate at each condition. For coronal domains whose solvency is better than Θ-solvent conditions, this results in a hard-core Yukawa-like repulsion for all cases studied. The range of this repulsion depends on the solvophilic block length, whereas its strength decreases with increasing solvophilic block chain due to a decrease of their effective grafting density, which leads to a weak dependence of the second virial coefficient on solvophilic block length.

We find that the phase stability of a dilute diblock copolymer micelle suspension is only weakly affected by the nature of the core (solvophobicity and chain length of the core blocks). Not surprisingly, colloidal suspensions of diblock copolymer micelles are always stable (normalised second virial coefficient $B_2^* > -6$) unless the solvophilic blocks are near Θ-solvent or in poor solvent conditions. For fixed core-forming block properties but different solvophilic block length the normalised second virial coefficients with varying coronal block solvency follow a similar curve. Furthermore, and contrary to what is expected from polymer-grafted colloidal particles, increasing the effective grafting density of solvophilic blocks from the micelle core *decreases* the phase stability of the micellar suspension. This is explained due to the interplay between the effective grafting density of solvophilic blocks from the core and the properties of the solvophilic and solvophobic blocks. The SCF method presented here for micelle–micelle interactions is extended to account for more components in solution in the next chapter.

References

1. F.A.M. Leermakers, J.C. Eriksson, J. Lyklema, Soft colloids, in *Fundamentals of Interface and Colloid Science*, vol. 5, ed. by J. Lyklema (Academic Press, Cambridge, 2005) Chap. 4, https://www.sciencedirect.com/bookseries/fundamentals-of-interface-and-colloid-science/vol/5/suppl/C
2. Y. Mai, A. Eisenberg, Chem. Soc. Rev. **41**, 5969 (2012), https://pubs.rsc.org/en/content/articlelanding/2012/cs/c2cs35115c#!divAbstract
3. A. Muñoz-Bonilla, S.I. Ali, A. del Campo, M. Fernández-García, A.M. van Herk, J.P.A. Heuts, Macromolecules **44**, 4282–4290 (2011), https://doi.org/10.1021/ma200626p
4. R. Tuinier, C.G. de Kruif, J. Chem. Phys. **117**, 1290 (2002), https://doi.org/10.1063/1.1484379?journalCode=jcp
5. L. Yang, X. Qi, P. Liu, A. El Ghzaoui, S. Li, Int. J. Pharm. **394**, 43 (2010), https://www.sciencedirect.com/science/article/pii/S0378517310003029
6. W. Li, J. Li, J. Gao, B. Li, Y. Xia, Y. Meng, Y. Yu, H. Chen, J. Dai, H. Wang, Y. Guo, Biomaterials **32**, 3832 (2011), https://www.sciencedirect.com/science/article/pii/S0142961211001219
7. J. Wang, X. Xing, X. Fang, C. Zhou, F. Huang, Z. Wu, J. Lou, W. Liang, Phil. Trans. R. Soc. A **371**, 20120309 (2013)
8. D. Lombardo, P. Calandra, D. Barreca, S. Magazù, M.A. Kiselev, Nanomaterials **6**, 125 (2016), https://www.mdpi.com/2079-4991/6/7/125/htm
9. M. Silbert, E. Canessa, M.J. Grimson, O.H. Scalise, J. Phys.: Condens. Matter **11**, 10119 (1999), https://doi.org/10.1088/0953-8984/11/50/306/meta
10. J.N. Israelachvili, *Intermolecular and Surface Forces*, 3rd edn. (Academic Press, Amsterdam, 2011)
11. I. Hamley, *Block Copolymers in Solution: Fundamentals and Applications* (Wiley, Hoboken, 2005)
12. E.B. Zhulina, O.V. Borisov, V.A. Priamitsyn, J. Colloid Interface Sci. **137**, 495 (1990), https://www.sciencedirect.com/science/article/pii/002197979090423L
13. E.K. Lin, A.P. Gast, Macromolecules **29**, 390 (1996), https://doi.org/10.1021/ma9505282
14. A.P. Gast, Langmuir **12**, 4060 (1996), https://doi.org/10.1021/la951538z
15. R. Lund, L. Willner, D. Richter, E.E. Dormidontova, Macromolecules **39**, 4566 (2006), https://doi.org/10.1021/ma060328y
16. S.-H. Choi, T.P. Lodge, F.S. Bates, Phys. Rev. Lett. **104**, 047802 (2010), https://doi.org/10.1103/PhysRevLett.104.047802
17. S.-H. Choi, F.S. Bates, T.P. Lodge, Macromolecules **44**, 3594 (2011), https://doi.org/10.1021/ma102788v
18. G.M. Grason, J. Chem. Phys. **126**, 114904 (2007), https://doi.org/10.1063/1.2709646

19. F. Puaud, T. Nicolai, E. Nicol, L. Benyahia, G. Brotons, Phys. Rev. Lett. **110** (2013), https://doi.org/10.1103/PhysRevLett.110.028302
20. C.N. Likos, H.Löwen, M. Watzlawek, B. Abbas, O. Jucknischke, J. Allgaier, D. Richter, Phys. Rev. Lett. **80**, 4450 (1998), https://doi.org/10.1103/PhysRevLett.80.4450
21. E.B. Zhulina, M. Adam, I. LaRue, S.S. Sheiko, M. Rubinstein, Macromolecules **38**, 5330 (2005), https://doi.org/10.1021/ma048102n
22. A.N. Semenov, I.A. Nyrkova, A.R. Khokhlov, Macromolecules **28**, 7491 (1995), https://doi.org/10.1021/ma00126a029
23. M.L. Kurnaz, J.V. Maher, Phys. Rev. E **55**, 572 (1997), https://doi.org/10.1103/PhysRevE.55.572
24. A. Quigley, D. Williams, Eur. J. Pharm. Biopharm. **96**, 282 (2015), https://www.sciencedirect.com/science/article/pii/S0939641115003288
25. Č. Koňák, Z. Tuzar, P. Štěpánek, B. Sedláček, P. Kratochvíl, *Frontiers in Polymer Science*, ed. by W. Wilke (Steinkopff, Darmstadt, 1985), pp. 15–19, in https://doi.org/10.1007/BFb0114009
26. M. Villacampa, E. Diaz de Apodaca, J.R. Quintana, I. Katime, Macromolecules **28**, 4144 (1995), https://doi.org/10.1021/ma00116a014#citing
27. T. Yoshimura, K. Esumi, J. Colloid Interface Sci. **276**, 450 (2004), https://doi.org/10.1016/j.jcis.2004.03.069
28. W. Li, M. Nakayama, J. Akimoto, T. Okano, Polymer **52**, 3783 (2011), https://doi.org/10.1016/j.polymer.2011.06.026
29. T. Zinn, L. Willner, R. Lund, V. Pipich, M.-S. Appavou, D. Richter, Soft Matter **10**, 5212 (2014), https://doi.org/10.1039/C4SM00625A
30. A. Mulero, *Theory and simulations of Hard-Sphere Fluids and Related System* (Springer, Heidelberg, 2008)
31. G.A. Vliegenthart, H.N.W. Lekkerkerker, J. Chem. Phys. **112**, 5364 (2000), https://doi.org/10.1063/1.481106
32. R. Tuinier, M.S. Feenstra, Langmuir **30**, 13121 (2014), https://doi.org/10.1021/la5023856
33. F. Platten, N.E. Valadez-Pérez, R. Castañeda-Priego, S.U. Egelhaaf, J. Chem. Phys. **142**, 174905 (2015), https://doi.org/10.1063/1.4919127
34. E.B. Zhulina, O.V. Borisov, Macromolecules **45**, 4429 (2012), https://doi.org/10.1021/ma300195n
35. E. Helfand, Y. Tagami, J. Polym. Sci., Part B: Polym. Lett. **9**, 741 (1971), https://doi.org/10.1002/pol.1971.110091006
36. Á. González García, A. Ianiro, R. Tuinier, ACS Omega (Supplemental information) **3**, 17976 (2018), https://doi.org/10.1021/acsomega.8b02548
37. H.N.W. Lekkerkerker, R. Tuinier, *Colloids and the Depletion Interaction* (Springer, Heidelberg, 2011)
38. J.M.H.M. Scheutjens, G.J. Fleer, J. Chem. Phys. **83**, 1619 (1979), https://doi.org/10.1021/j100475a012
39. J.M.H.M. Scheutjens, G.J. Fleer, J. Chem. Phys. **84**, 178 (1980), https://doi.org/10.1021/j100439a011
40. G.J. Fleer, M.A. Cohen Stuart, J.M.H.M. Scheutjens, T. Cosgrove, B. Vincent, *Polymers at Interfaces* (Springer, Netherlands, 1998), pp. XX, 496
41. C.B.E. Guerin, I. Szleifer, Langmuir **15**, 7901 (1999)
42. C. Guerrero-Sanchez, D. Wouters, C.-A. Fustin, J.-F. Gohy, B.G.G. Lohmeijer, U.S. Schubert, Macromolecules **38**, 10185 (2005), https://doi.org/10.1021/ma051544u
43. M. Amann, L. Willner, J. Stellbrink, A. Radulescu, D. Richter, Soft Matter **11**, 4208 (2015), https://doi.org/10.1039/C5SM00469A
44. K. van Gruijthuijsen, M. Obiols-Rabasa, M. Heinen, G. Nägele, A. Stradner, Langmuir **29**, 11199 (2013), https://doi.org/10.1021/la402104q
45. C.N. Likos, H.M. Harreis, Condens. Matter Phys. **5**, 173 (2002), http://dspace.nbuv.gov.ua/handle/123456789/120565
46. S.-H. Chen, M. Broccio, Y. Liu, E. Fratini, P. Baglioni, J. Appl. Crystallogr. **40**, s321 (2007), https://doi.org/10.1107/S0021889807006723

47. G.J. Brown, R.W. Richards, R.K. Heenan, Polymer **42**, 7663 (2001), https://www.sciencedirect.com/science/article/pii/S003238610100252X
48. J. Bergsma, F.A.M. Leermakers, J. van der Gucht, Phys. Chem. Chem. Phys. **17**, 9001 (2015), https://doi.org/10.1039/C4CP03508A
49. E.B. Zhulina, O.V. Borisov, Makromol. Chem., Macromol. Symp. **44**, 275 (1991), https://doi.org/10.1002/masy.19910440128
50. D.N. Benoit, H. Zhu, M.H. Lilierose, R.A. Verm, N. Ali, A.N. Morrison, J.D. Fortner, C. Avendano, V.L. Colvin, Anal. Chem. **84**, 9238 (2012), https://doi.org/10.1021/ac301980a

Chapter 8
Polymer-Mediated Stability of Micellar Suspensions

8.1 Introduction

Association colloids are formed from amphipathic building blocks called unimers [1]. Often, the properties of the self-assembled structure are determined by its equilibrium with the bulk unimer concentration. The unimer nature and the solvency conditions does not only affect the preferred morphology of the self-assembled structures [2], but also their colloidal stability [3]. Spherical micelles formed from block copolymers in solution have received ample interest for multiple applications, including drug delivery [4–7], coatings [8], and are present in foodstuffs [9, 10]. Diblock copolymers are constituted by a solvophilic and a solvophobic block, consisting of m-solvophobic and n-solvophilic units respectively. The solvophobic blocks drive the micelle formation, as solvophobic segments (B) tend to minimize their contact with the solvent by forming a compact, solvent-depleted core [11]. The solvophilic blocks (A) concentrate in the well-solvated corona. From both fundamental and application perspectives, biocompatible neutral block copolymers are appealing: then the system parameters are narrowed down to the length, nature, and sequence of the blocks. Potentially, this provides better control over the system of interest [12]. In Chap. 7, we showed that whereas the equilibrium properties of micelles are dominated by the relative block lengths and block solvencies, the colloidal stability of a micelle suspension is mainly determined by the solvency of the coronal block.

Commonly, micelles are not the only component in solution, and the inherent steric stabilisation between micelles [12] may get compromised due to depletion or adsorption of other compounds in solution. For instance, a micellar drug delivery system acts in the presence of a myriad of components that may alter the micellar properties [13] and affect its colloidal stability [3]. The surface of a hard colloidal sphere is sharp and impenetrable for free polymers in solution [14]. Differently, the peripheral region of a micelle is diffuse (Fig. 8.1), hence partial penetration of the homopolymers may occur [15]. Additionally, the equilibrium properties of micelles, due to their associative nature, are influenced by the presence of added compounds (see Fig. 8.1). The capability of polymers to penetrate the micellar domain depends

© Springer Nature Switzerland AG 2019

Á. González García, *Polymer-Mediated Phase Stability of Colloids*,
Springer Theses, https://doi.org/10.1007/978-3-030-33683-7_8

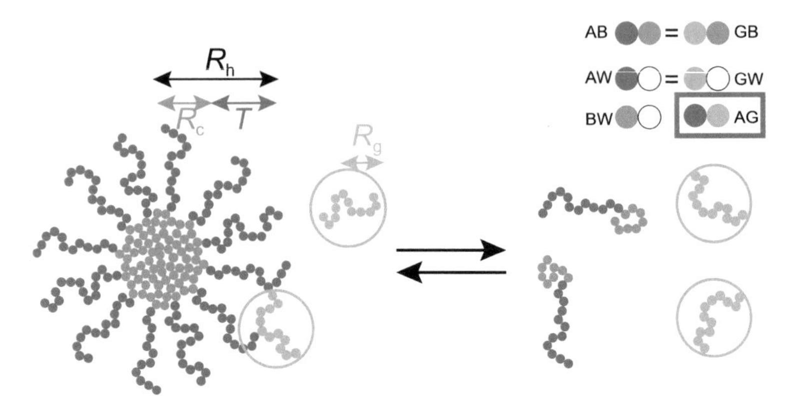

Fig. 8.1 Representation of micelle–unimer equilibrium of a micelle (left) composed of g_p unimers in the presence of added homopolymer. Addition of a second component in the bulk (potentially) affects this equilibrium. The top-right quadrant corresponds to the segment–segment and segment–solvent interactions considered (values given in Table 8.1), with A the solvophilic block, B the solvophobic block, G the homopolymer block, and W a solvent molecule. Further specified are the hydrodynamic micelle radius R_h, the corona thickness T, the core radius R_c, and the radius of gyration of the homopolymer R_g. We highlight the parameter χ_{AG} which determines whether homopolymers are depleted or adsorbed

on the local density of the corona and on the interaction between the free polymer and the corona-forming blocks.

Even though widely present in biology and man-made products, there is yet limited fundamental understanding of the effect of added homopolymer into a suspension of spherical micelles [16], and most of the investigations focused on the solid phases of micelles and how these are influenced by homopolymer addition [17, 18]. The model presented in Chap. 7 is extended here to account for addition of homopolymers to the micellar suspension. Often, a distinction is made between crew-cut (similar core and corona sizes) and starlike (significantly larger corona than core) micelles [19, 20]. We investigate how the presence of homopolymer in solution affects the micelle–unimer equilibrium both for crew-cut and starlike micelles.

From the collected polymer-mediated micelle–micelle interactions, we calculated the second virial coefficient B_2 [21, 22] to assess the phase stability of association colloid–polymer mixtures (ACPMs). Consider a dilute micelle suspension at a fixed diblock concentration (above the critical micelle concentration, CMC) to which guest compounds are added. If such a mixture has a B_2-value near or above the hard sphere limit [23], it is expected to remain optically transparent ($B_2^* \equiv B_2/v_c = 4$, with v_c the considered colloidal particle volume). If instead micelles (strongly) attract each other, $B_2^* < 4$ and the micelle suspension may get turbid. In case of $B_2^* \leq -6$, the Vliegenthart–Lekkerkeker (VL) criterion [24] states that colloidal gas–liquid phase separation of the micellar dispersion is expected (see discussion in Sect. 1.3.2). Even though this B_2-value is in principle experimentally resolvable [25–29], there are many practical difficulties related to measuring B_2. Thus, direct visual observation of the (in)stability of the ACPM serves here as a pragmatic confirmation of theoretical predictions made. Trends found based on our theoretical model are compared with the stability of polycaprolactone–polyethylene glycol (PCL-PEO) micelles in water in the presence of added PEO chains. The results reported show why micelles with a narrow corona are more suitable for micellar applications in crowded environments.

8.2 System Parameters

We followed the approach explained in Sect. 1.3.5 to collect (homo)polymer-mediated micelle–micelle interactions. When considering diblock copolymer micelles in the presence of polymers in a common solvent, the key variables are those reported in Fig. 8.1. Specifically, these variables are: (i) the different segment–segment and segment–solvent interactions; (ii) the equilibrium micelle properties, and (iii) the relative size of homopolymer to micelle and its bulk concentration. Micelles are constituted of g_{p} (the aggregation number) diblock copolymers with composition $\mathrm{B}_{24}\mathrm{A}_n$, where B is the solvophobic block ($\chi_{\mathrm{BW}} = 2$), A is the solvophilic block ($\chi_{\mathrm{AW}} = 0.4$), and $\chi_{\mathrm{AB}} = 1$. These parameters are selected to meet typical solvencies for industrially relevant systems such as pluronics in water [30, 31], and are based on previous investigations [32]. At fixed solvophobic block length ($m = 24$) we consider a crew-cut and a starlike micelle with $n = 45$ and $n = 450$, respectively. In Table 8.1 the parameters that define the architecture of the diblocks and guest polymers are specified.

To assess the effects of the homopolymer-mediated micelle–micelle interactions, we introduce the size ratio of the (guest) homopolymer to the undistorted micelle size:

$$q \equiv \frac{R_{\mathrm{g}}}{R_{\mathrm{h}}^{\mathrm{o}}}, \tag{8.1}$$

where $R_{\mathrm{h}}^{\mathrm{o}}$ is the hydrodynamic radius of an isolated micelle in absence of homopolymer (see Chap. 1 for details on the calculation of R_{h}). The radius of gyration of the free homopolymer composed of N segments in solution is approximated as [33]:

Table 8.1 System parameters chosen in this study: number of solvophobic (m), solvophilic (n), and guest polymer segment (N) repeating units, and Flory–Huggins interaction parameters. A and B refer to a solvophilic and solvophobic segment respectively; G denotes a guest homopolymer segment; and W is a solvent molecule

m	n	N	χ_{BW}	$\chi_{\mathrm{AW}} \equiv \chi_{\mathrm{GW}}$	$\chi_{\mathrm{AB}} \equiv \chi_{\mathrm{GB}}$	χ_{AG}
24	45, 450	var	2	0.4	1	var

$$R_g = 0.309bN^{1/2}\left[1+\sqrt{1+6.5N^{1/2}(1-2\chi_{GW})}\right]^{0.352}, \tag{8.2}$$

where b is the monomer size which equals the size of a lattice site. The experimentally-resolvable value for R_h^o qualitatively agrees with theoretical predictions for diblock copolymer micelle systems [34]. Though approximated, Eq. (8.2) captures the SCF-predicted trends for the radius of gyration of an isolated guest homopolymer chain. The q-value is commonly used to rationalise depletion phenomena [33]. Provided R_h^o is known, the size of the homopolymer can be obtained via Eqs. (8.1) and (8.2) for a certain (imposed) q-value. Finally, we conducted our computations at fixed homopolymer bulk concentration ϕ_G^{bulk}. The homopolymer concentration relative to overlap is defined as:

$$\phi_G^* = \frac{Nb^3}{v_G} \tag{8.3}$$

with

$$v_G = \frac{4\pi}{3}R_g^3 \tag{8.4}$$

the estimated volume occupied by a (guest) homopolymer coil in the bulk solution. The guest (G) polymer is considered to be in the same solvent condition as the solvophilic block segments ($\chi_{GW} \equiv \chi_{AW} = 0.4$), and the solvophobic segment–homopolymer interaction is considered as the one between A and B segments ($\chi_{GB} \equiv \chi_{AB} = 1$). The effective affinity of the guest compound to the corona of the micelle is varied through χ_{AG}, where $\chi_{AG} = 0$ corresponds to athermal (excluded volume) corona–homopolymer interactions.

8.3 Results and Discussion

We first discuss the equilibrium properties of micelles and the concentration distribution of a dilute homopolymer solution as a function of the homopolymer–corona effective affinity. Subsequently, homopolymer concentration effects on the micelle equilibrium are studied for different guest polymer–micelle combinations (crew-cut, starlike, depleted, adsorbed). Representative homopolymer-mediated micelle–micelle interactions are rationalised based on the micelle–unimer equilibrium shifts upon addition of homopolymer. Additionally, the stability of association colloid–polymer mixtures (ACPMs) is investigated, and results are discussed in terms of the second virial coefficient B_2.

Table 8.2 Equilibrium properties of the example crew-cut and starlike micelles considered: aggregation number g_p, critical micelle concentration CMC (ϕ_p^{bulk}), maximum concentration of solvophilic blocks ϕ_A^{max}, undistorted hydrodynamic size R_h^o, core radius R_c, and coronal thickness T. Dimensions are expressed in lattice units [l.u.]

Type	Unimer	g_p	$\phi_p^{bulk} \approx$ CMC	ϕ_A^{max}	R_h^o	R_c	T
Crew-cut	$B_{24}A_{45}$	134	3.6×10^{-8}	0.31	19.5	11	8.5
Starlike	$B_{24}A_{450}$	37	1.3×10^{-5}	0.20	48.3	8	40.3

8.3.1 Dilute Homopolymer in a Micellar Suspension

The SCF-computed characteristics of isolated crew-cut and starlike micelles are presented in Table 8.2. The different architecture of the unimers leads to distinct properties of the assemblies, reflected in their aggregation number g_p, undistorted hydrodynamic size R_h^o, critical micelle concentration (CMC), and sharpness of the corona (expressed through the maximum concentration of solvophilic blocks ϕ_A^{max}). Due to the (much) lower g_p-value but larger R_h^o-value of the starlike micelle compared to the crew-cut one, the available surface area per diblock of a starlike micelle is significantly larger than for a crew-cut one. Hence, the peripheral region of the starlike micelle is more diffuse than for the crew-cut one. Due to the larger solvophilic to solvophobic block size ratio of the starlike micelle, its CMC (expressed here via the diblock bulk volume fraction ϕ_p^{bulk}) is about 360 times larger. Consequently, the unimer–micelle equilibrium upon addition of guest homopolymer is expected to be greatly affected for a starlike micelle. Equilibrium concentration profiles are shown on the top panels of Fig. 8.2 as a function of the distance from the centre of the micelle z. The maximum concentration of solvophilic blocks ϕ_A^{max} roughly coincides with the position of the core–corona interface, used here as estimated core radius [$R_c \equiv z(\phi_A^{max})$]. The corona thickness (defined as $T = R_h^o - R_c$) is obviously larger for the starlike micelle, which has a smaller R_c due to the much lower value of g_p. The lower ϕ_A^{max}-value also indicates that the outer region of the starlike micelle is overall less dense than for the crew-cut assembly.

Not only the relative size of the homopolymer and its concentration modulate the effective micelle–micelle interactions, but also the affinity between the homopolymer and the colloidal surface [35]. Homopolymers in solution are either depleted from or adsorbed to the corona domain, depending on the affinity between the corona blocks and the homopolymer. In the bottom panels of Fig. 8.2, the normalised homopolymer segment concentration (ϕ_G/ϕ_G^{bulk}) is plotted as a function of z at fixed $\phi_G^{bulk}/\phi_G^* = 10^{-4}$. We define here the adsorption thickness from the discrete SCF density profile $\phi_G(z)$ as:

$$\frac{\delta}{b} = \sum_{z=z_0}^{N_{lat}} \frac{\phi_G(z)}{\phi_G^{bulk}} - 1, \tag{8.5}$$

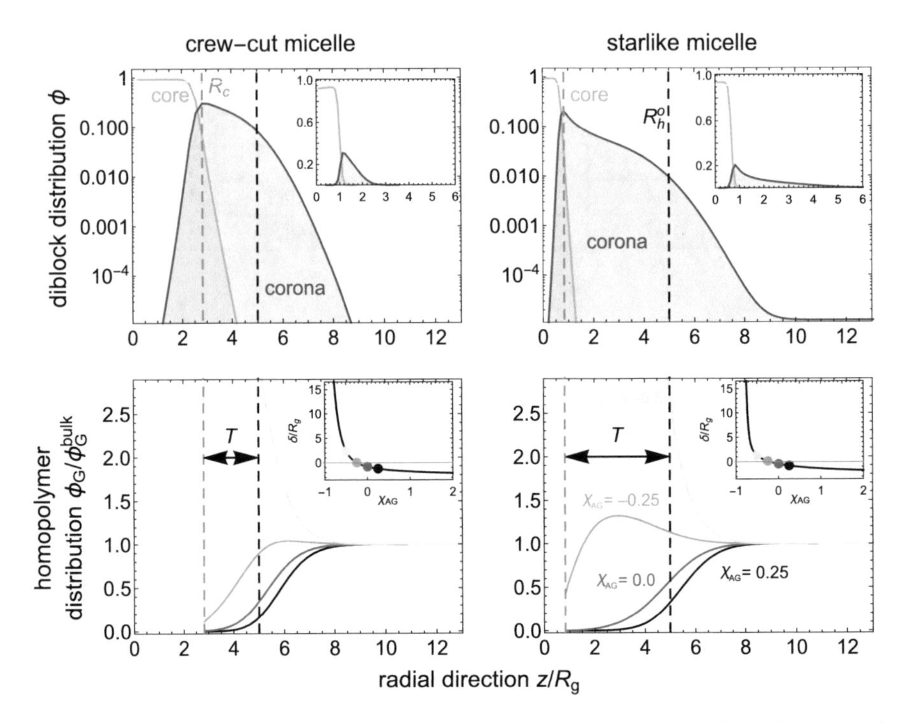

Fig. 8.2 Top panels: micelle-forming diblock segment concentration profiles from the centre of a micelle, computed using SCF theory in a spherical lattice considering a crew-cut ($B_{24}A_{45}$, left panels) and a starlike ($B_{24}A_{450}$, right panels) micelle. Core and corona regions are indicated; dashed lines indicate the core radius (R_c, grey) and the calculated hydrodynamic radius (R_h, black). Insets as main plots, but in a linear scale. Bottom panels: homopolymer concentration profiles relative to the bulk concentration. All results refer to calculations for a dilute homopolymer solution ($\phi_G^{bulk}/\phi_G^* = 10^{-4}$) with relative size $q \equiv R_g/R_h^o = 0.2$. In the insets, the adsorption thickness δ as a function of the corona–homopolymer effective affinity χ_{AG} is plotted. The χ_{AG}-values used in the main plot are indicated with discs: $\{0.25, 0, -0.25, -0.5\}$

where z_0 is the considered offset layer ($z = \{1, 2, ..., N_{lat}\}$, with N_{lat} the lattice size considered). For comparison with depletion and adsorption between hard surfaces, this offset layer was chosen as the closest layer to R_h^o; $z_0 \approx R_h^o$. A negative δ-value is associated with depletion phenomena, whilst $\delta > 0$ is indicative of homopolymer adsorption. For $\chi_{AG} \geq 0$ homopolymers are *depleted* ($\phi_G < \phi_G^{bulk}$) from the micelle, with $\phi_G(z = R_c) \approx 0$. Due to the entropic penalty of homopolymers being near or within the micelle, ϕ_G decreases with respect to ϕ_G^{bulk} if χ_{AG} attains a sufficiently large value.

For $\chi_{AG} = -0.25$, $\phi_G(z \lesssim R_h^o)$ is larger than for $\chi_{AG} = 0.25$ or $\chi_{AG} = 0$. For the crew-cut micelle the homopolymer practically does not adsorb at $z > R_h^o$ for $\chi_{AG} = -0.25$, and is still partially depleted from R_c (note there is a finite homopolymer concentration at R_c). For the starlike micelle, ϕ_G reaches a maximum value above ϕ_G^{bulk} within the corona; yet it is also partially depleted close to the core ($z \to R_c$).

This suggests that the degree of penetration of the added polymer into the coronal domains modulates the stability of ACPMs. For $\chi_{AG} = -0.5$, ϕ_G is always greater than ϕ_G^{bulk} for z-values within the corona: homopolymer adsorption to the corona occurs.

The transition from homopolymer depletion to adsorption is observed in the insets of the bottom panels of Fig. 8.2 in terms of δ as a function of χ_{AG}. Around $\chi_{AG} = -0.25$, the sign of δ switches from negative (depletion) to positive (adsorption). The thickness of the depletion layer $|\delta|$ increases with increasing χ_{AG} until it reaches a plateau, in concordance with what is expected for depletion from a hard surface. The dramatic increase in δ upon homopolymer adsorption is also in line with observations of homopolymer adsorption at a hard surface [14].

8.3.2 Depletion in ACPMs

Firstly, we focus on cases where depletion of the homopolymer is observed in Fig. 8.2. Particularly, we study the effect of homopolymer concentration ϕ_G^{bulk} on the micelle–unimer equilibrium and on the local density profiles near and within an isolated micelle. In Fig. 8.3 the variation in aggregation number, expressed in terms of

$$\Delta g_p = g_p(\phi_G^{bulk}) - g_p(\phi_G^{bulk} = 0), \tag{8.6}$$

is shown for different depletion cases. As observed, with increasing ϕ_G^{bulk} the aggregation number of both kinds of micelles increases. This agrees with previously reported

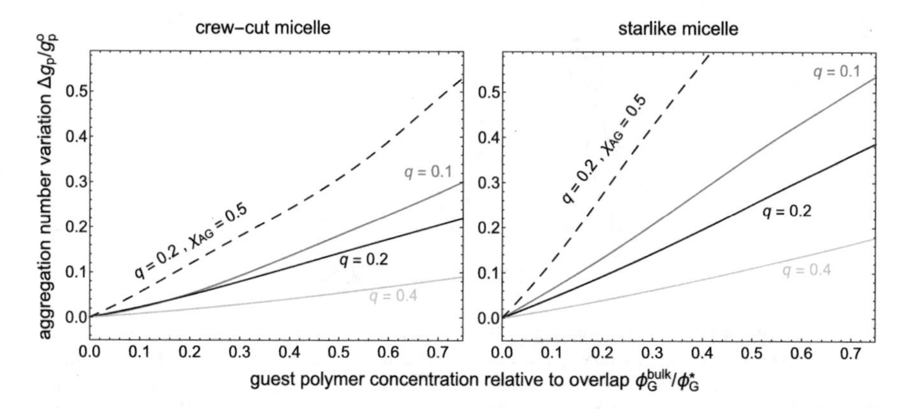

Fig. 8.3 Equilibrium aggregation number variation Δg_p of the considered crew-cut (right panel) and starlike (left panel) micelles with increasing homopolymer bulk concentration ϕ_G^{bulk}. Solid curves correspond to depleted homopolymers with athermal interactions with the corona ($\chi_{AG} = 0$). Homopolymer–to–micelle size ratio q varied as indicated. Dashed curves correspond to a homopolymer with more repulsive interaction with the coronal blocks ($\chi_{AG} = 0.5$)

experimental observations [16]. The increase in g_p is weaker when the homopolymer size is larger (i.e., increasing q). The reason for this is two-fold. On the one hand, the un-normalised ϕ_G^{bulk}-value decreases with increasing q [see Eq. (8.3)]. On the other hand, larger q-values imply that less homopolymers can penetrate into the coronal domain. Interestingly, the micelle size is hardly affected by these changes in g_p. In fact, upon averaging over all considered q-values and ϕ_G^{bulk} concentrations in Fig. 8.3, we find for the crew cut micelle $\langle R_h \rangle = 19.8 \pm 0.2$ [l.u.] and for the starlike micelle $\langle R_h \rangle = 48.0 \pm 0.9$ [l.u.]. These values are fairly close to the depletant-free, undistorted micelle sizes presented (see Table 8.2). This increase on g_p relates to unfavourable interactions between the homopolymers and the diblocks in bulk. Addition of homopolymer to the bulk shifts the micelle–unimer equilibria towards the micelle, thus increasing g_p. This is confirmed by the fact that for $q = 0.2$ and $\chi_{AG} = 0.5$, g_p is larger than for $\chi_{AG} = 0$ at the same ϕ_G^{bulk}. The steeper increase of g_p for the starlike micelle follows from its (much) higher ϕ_p^{bulk} (see Table 8.2).

Homopolymer segment concentration distributions for selected ϕ_G^{bulk}-values are shown in Fig. 8.4. The observed partial penetration of homopolymers in the coronal domain has been previously reported [17], and such interpenetration also occurs between polymer brushes and free homopolymer [15, 36]. With increasing homopolymer bulk concentration, $|\delta|$ decreases due to the increasing osmotic pressure that bulk homopolymers exert onto those in the vicinity or within the diffuse micelle's peripheral region. This effect is well-known for non-adsorbing polymers near a hard surface [37, 38]. As depletants penetrate through the coronal region,

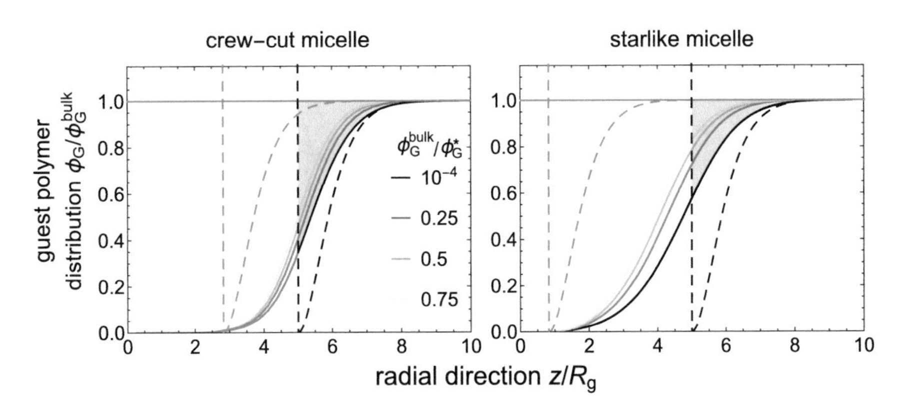

Fig. 8.4 Homopolymer (depletant) segment concentration profiles as a function of the distance from the centre of a micelle z as computed using SCF theory in a spherical lattice considering crew-cut (left panel) and starlike (right panel) micelles. The homopolymer considered is of the same nature as the corona ($\chi_{AG} = 0$). Dashed curves correspond to the depletion profiles of polymers in Θ-solvent condition from a hard sphere [33], considering the hard sphere radius either as the core radius (R_c, grey) or the hydrodynamic radius of the undistorted micelle (R_h^o, black). The relative polymer size is $q \equiv R_g/R_h^o = 0.2$. Results are given for the indicated homopolymer concentrations ϕ_G^{bulk}. The coloured grey areas are used to illustrate the depletion thickness δ from $z_0 \approx R_h^o$ for the lowest homopolymer concentration

this compression of the depletion layer is weaker for a micelle than for a hard colloidal surface. As can be observed, the shape of the depletion profile resembles that of non-adsorbing homopolymers in Θ-solvent around a hard sphere [33, 39] [$\tanh^2 (z - z_0) / |\delta|$]. However, in this case the depletion density profile extends in between the limits of a 'classical' depletion profile from hard spheres with radii R_h^o and R_c (dashed vertical lines in Fig. 8.4). A broader coronal thickness leads to a wider depletion profile for the starlike micelles.

In Fig. 8.5 the micelle–micelle interactions in presence of depleted homopolymers are presented. At low depletant concentrations ($\phi_G^{bulk}/\phi_G^* = 0.01$), the micelle–micelle interactions can be described as a hard-core Yukawa repulsion (see Chap. 7). At higher ϕ_G^{bulk}, pure excluded volume interactions of the homopolymer with the coronal blocks ($\chi_{AG} = 0$) induce a shallow minimum in the micelle–micelle interaction between crew-cut micelles. This minimum is due to the homopolymer-induced

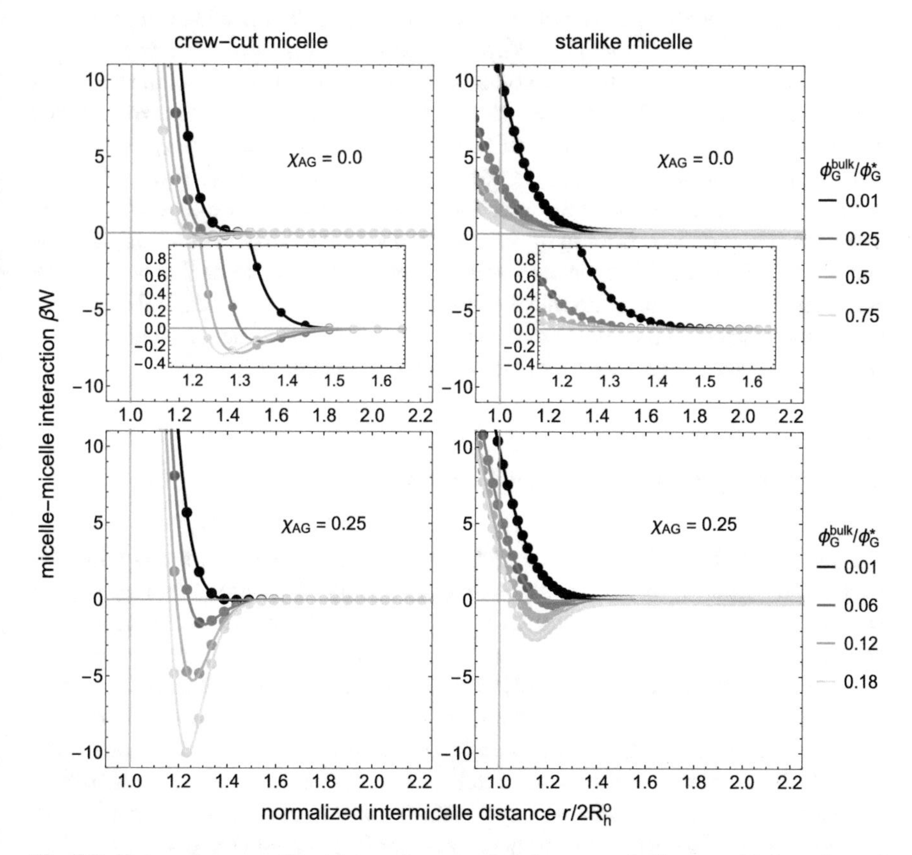

Fig. 8.5 Homopolymer-mediated interaction potentials between micelles, considering athermal homopolymer–corona interactions ($\chi_{AG} = 0$, top panels) and an additional repulsion ($\chi_{AG} = 0.25$, bottom panels) for the relative homopolymer concentrations ϕ_G^{bulk}/ϕ_G^* indicated. The relative polymer size is $q \equiv R_g/R_h^o = 0.2$

depletion attraction between micelles. The position of this minimum shifts towards $r = 2R_{\mathrm{h}}^{\mathrm{o}}$ with increasing $\phi_{\mathrm{G}}^{\mathrm{bulk}}$ due to the compression of the depletion layer (inset, left panel). For the starlike micelle considered, $\chi_{\mathrm{AG}} = 0$ is insufficiently repulsive to induce an attractive minimum in the interaction potential between the micelles (inset, right panel); the micelle–micelle interaction is only repulsive. However, this micelle–micelle repulsion significantly weakens in $r < 2R_{\mathrm{h}}^{\mathrm{o}}$ with increasing $\phi_{\mathrm{G}}^{\mathrm{bulk}}$. These effects are further rationalised in Sect. 8.3.4. For $\chi_{\mathrm{AG}} = 0.25$, a minimum in the homopolymer-mediated micelle–micelle interactions is present for the two kinds of micelles as a result of more repulsive corona–homopolymer interactions, which effectively shifts the depletion zone towards the outer micelle region: there is not only an entropic, but also an enthalpic penalty whenever depletants penetrate into the coronal domain.

Experimentally, depletion effects decrease the effective (hard-sphere equivalent) colloidal size of the micelle, which may relate to the lack of a repulsive contribution to the micelle–micelle interaction observed for the starlike micelle at high enough homopolymer concentrations [16]. We note here that whereas $|\delta|$ decreases with $\phi_{\mathrm{G}}^{\mathrm{bulk}}$, this decrease is not enough (in hard colloidal suspensions) to re-stabilize the depletion attraction, whose strength increases with $\phi_{\mathrm{G}}^{\mathrm{bulk}}$. The overlap of depletion zones when micelles get closer does not lead to full depletion of the guest homopolymer, even if an added enthalpic penalty is considered (profile details in Sect. 8.A). From the trends of the depletion profiles and of the micelle–micelle interactions, we conclude that the depletion attraction is weaker in micellar suspensions than in hard colloidal systems due to the inherent steric repulsion and the penetration of the depletants into the colloidal domain. Insights into micelle–micelle interactions may be of relevance for further understanding the effects of depletion phenomena in systems where micelles are present, such as drug-delivery systems [7] or foodstuffs [9].

8.3.3 *Coronal Physisorption in ACPMs*

First, we study the effect of corona-adsorbing homopolymers on the equilibrium micelle properties at fixed (homo)poloymer-to-micelle size ratio q. In Fig. 8.6, the variation of g_{p} due to polymers with an enthalpic preference for the corona is reported, and compared to guest polymers with athermal interaction with the corona. For isolated micelles and low guest polymer concentrations, adsorption to the corona of the starlike micelle is much higher than to the crew-cut one (detailed profile in Sect. 8.A). Contrary to the depletion case, the micelle–unimer equilibrium shifts towards the unimer side due to favourable diblock–homopolymer interactions in the bulk. When a sufficiently strong corona–homopolymer affinity is considered, homopolymers adsorbed at the core–corona interface screen the solvophobic core [40]. These two arguments suffice for explaining the decrease in g_{p} with increasing $\phi_{\mathrm{G}}^{\mathrm{bulk}}$ observed for crew-cut micelles, show in Fig. 8.6.

The effect of adsorption onto the corona of starlike micelles is more intricate due to the more diffuse peripheral domain. A polymer with a strong affinity for the (large)

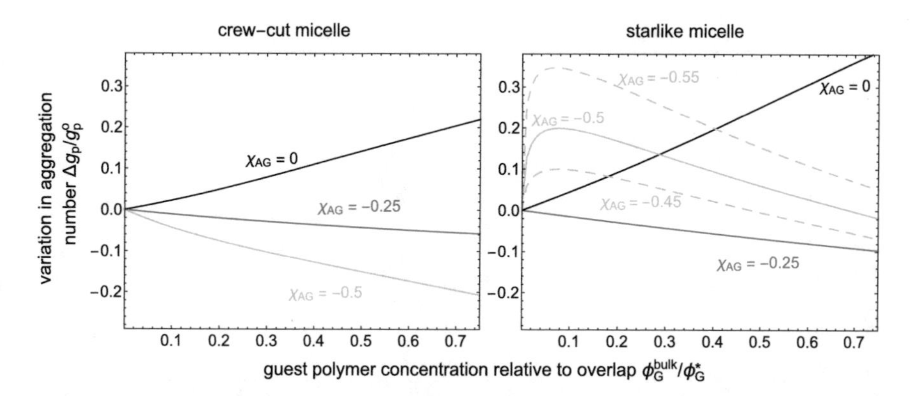

Fig. 8.6 Equilibrium aggregation number variation Δg_p of the considered crew-cut (right panel) and starlike (left panel) micelles with increasing homopolymer bulk concentration ϕ_G^{bulk}. The homopolymer–corona affinity χ_{AG} is indicated. Relative homopolymer-to-micelle size ratio is $q = R_g/R_h^o = 0.2$

corona may actually stick solvophilic brushes together, hence overcompensating the entropic penalty associated with the interpenetration of the homopolymers in the corona. This leads to the observed increase of g_p at very low ϕ_G^{bulk}. At low ϕ_G^{bulk} the micelle–unimer equilibrium displaces to the micelle: there is a strong preference of the homopolymer for the corona (g_p increases). Saturation of the corona with homopolymers then leads to a decrease of g_p at higher ϕ_G^{bulk}, hence the micelle–unimer equilibrium shifts towards the unimer state: homopolymer–diblock contacts in the bulk become, again, more favourable than in the saturated micelle. The curves for $\chi_{AG} = -0.45$ and $\chi_{AG} = -0.55$ for the starlike micelle corroborate this explanation. The same effect on g_p could be observed for stronger attractions ($\chi_{AG} \approx -0.7$) of the homopolymer with the coronal domain of the crew-cut micelle. From our theoretical investigations [see also [40] for encapsulation of small ($q \ll 0.1$) guest compounds], it is clear that the changes on g_p and R_h can be used to experimentally assess the distribution of adsorbing compounds over micelles.

In Fig. 8.7, example micelle–micelle interactions mediated by corona-adsorbing homopolymers are shown. For crew-cut micelles, a slight preference for the coronal blocks ($\chi_{AG} = -0.25$) hardly affects the micelle–micelle interactions upon increasing ϕ_G^{bulk}; even close to ϕ_G^*, the micelle–micelle interaction mostly remains unaffected. For a stronger homopolymer–corona affinity ($\chi_{AG} = -0.5$), a shallow attraction at low ϕ_G^{bulk}-values appears, which then weakens with increasing ϕ_G^{bulk} (inset of Fig. 8.7). This can be explained by a bridging attraction mechanism [14]: at low concentration homopolymers simultaneously adsorb onto the coronas of different micelles, which leads to attraction between them. This trend of the bridging attraction between micelles with increasing ϕ_G^{bulk} is similar to that observed between hard colloids (see Chap. 4). With increasing ϕ_G^{bulk}, the coronal domains get saturated with homopolymer, leading to a restabilisation of the micelle solution.

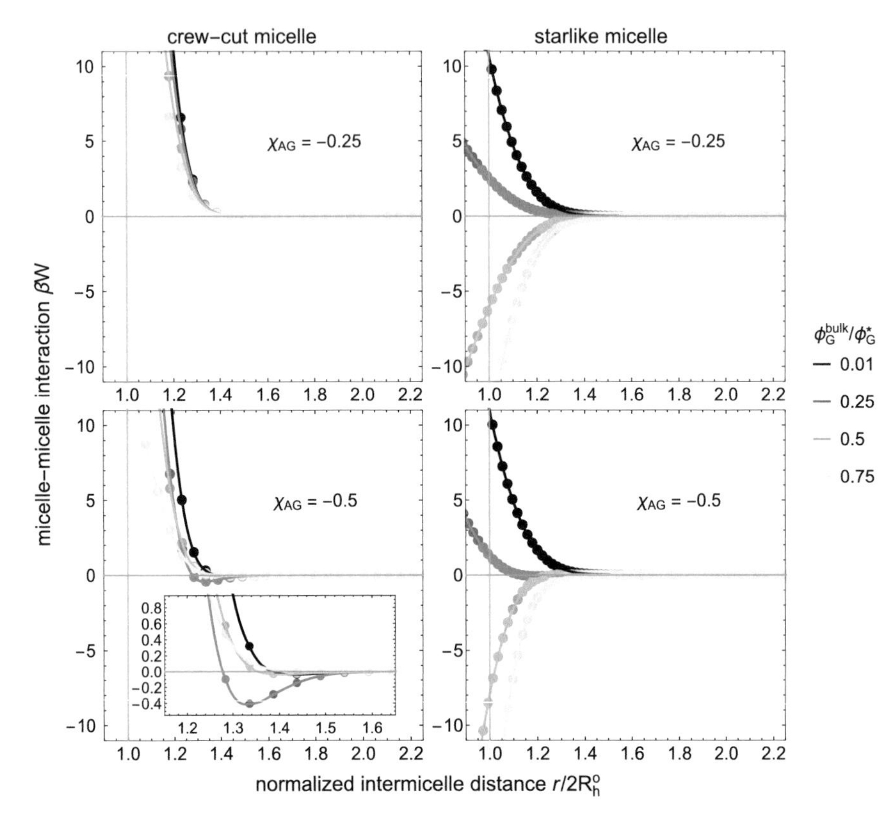

Fig. 8.7 Homopolymer-mediated interaction potentials between micelles, considering attractive homopolymer-corona interactions: $\chi_{AG} = -0.25$, top panels; and $\chi_{AG} = -0.5$, bottom panels. Homopolymer concentrations relative to overlap $\phi_G^{bulk}/\phi_G{*}$ are indicated. The relative polymer size is $q \equiv R_g/R_h^o = 0.2$

As for the depletion cases, the effect of adsorbing polymer is more convoluted for starlike micelles due to their broader corona. For the q-value considered ($q = 0.2$), the relative volume of the (undistorted) homopolymer per coronal block is about *four times* larger for the starlike micelle than for the crew-cut one (see Sect. 8.A). Note that at fixed q and χ_{AG}, adsorption is much higher for the starlike micelle than for the crew-cut one (see profiles in Sect. 8.A). Hence, bridging effects are strong for starlike micelles as they get closer. The large and fluffy corona prevents saturation of the peripheral colloidal domain, at least within the considered ϕ_G-values. Thus, contrary to the crew-cut micelle, re-stabilisation of the bridging flocculation is not observed in the micelle–micelle interactions even at high ϕ_G. The steric micelle–micelle repulsion gets screened due to the presence of adsorbing homopolymer, leading to a transition from repulsive to attractive micelle–micelle interaction. In the next section, the relevance of the relative size of the added homopolymer to the coronal thickness is addressed: penetration of homopolymer is rationalised not in terms of $q \equiv R_g/R_h^o$, but in terms of R_g/T.

8.3.4 Corona Thickness and Colloidal Stability

It is clear from the computed micelle–micelle interactions that a high degree of interpenetration of either adsorbing or depleted compounds into the coronal domain leads to destabilisation of the micelle suspension with increasing guest homopolymer concentration ϕ_G^{bulk}. For the relative size $q = 0.2$, the starlike micelle–micelle steric repulsion practically vanishes at $\phi_G^{bulk}/\phi_G^* = 0.75$ considering corona-like homopolymers ($\chi_{AG} = 0$, see Fig. 8.5, top right panel). Within the model here considered, the strongest repulsive contribution to the (homopolymer-free) micelle–micelle interaction arises from the expel of diblocks (core compression, see Chap. 7). Independently of the value of χ_{AG}, the presence of homopolymers in the corona may screen this strongly-repulsive core compression, leading to micelle–micelle attractions even for $r < 2R_h^o$. For the crew-cut micelle considered, $q \approx 0.2$ corresponds to the situation where the diameter of the added homopolymer roughly equals the coronal thickness ($2R_g = T$). For a starlike micelle, this situation is retained for $q \approx 0.4$.

We present in Fig. 8.8 the homopolymer-mediated starlike micelle–micelle interaction for $q = 0.4$ and $\chi_{AG} = 0$. These micelle–micelle interactions are similar to those between crew-cut micelles at $q = 0.2$ (for which $R_g = T/2$): for larger q, fewer polymer chains fit into the corona, and a repulsive contribution to the homopolymer-mediated micelle–micelle interaction remains. The much higher ϕ_G^{bulk}-value for the starlike micelle renders all guest homopolymer effects on the micelle equilibria stronger due to the lower energy required to remove a diblock from the starlike micelle as compared to the crew-cut one. This leads to a clear decrease of the repulsive contribution to the micelle–micelle interaction, which decreases from $W(r = 2R_h^o) \approx 10k_BT$ in absence of homopolymer to $W(r = 2R_h^o) \approx 5k_BT$ at $\phi_G^{bulk}/\phi_G^* = 0.75$. It follows from this short section that it is the ratio of the corona thickness to the diameter of the added homopolymer which determines whether a repulsive contribution to the micelle–micelle interaction is present upon addition of homopolymer. Therefore, the ratio T/R_g might be of relevance in the design of controlled experiments.

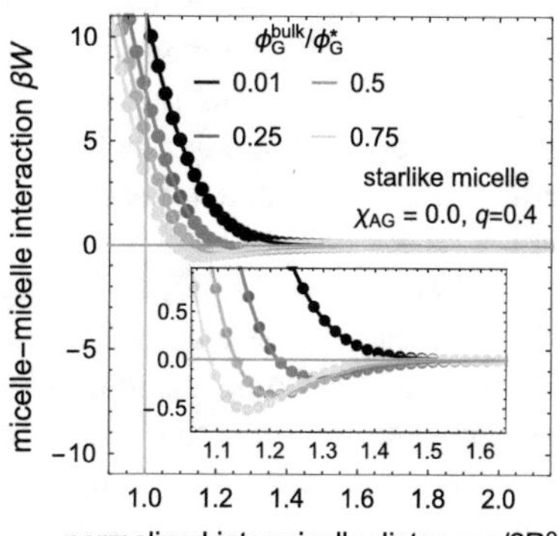

Fig. 8.8 Homopolymer-mediated interaction potentials between micelles, considering athermal homopolymer-corona interactions ($\chi_{AG} = 0$) for starlike micelles in presence of homopolymer with relative size $q = 0.4$

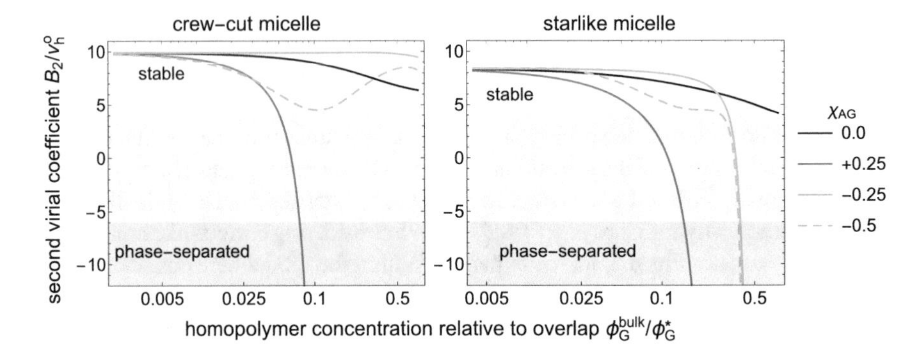

Fig. 8.9 Second virial coefficient normalised by the undistorted micelle volume for $q \equiv R_g/R_h^o = 0.2$ for crew-cut (left) and starlike (right) panel. The effective affinity between the homopolymer and the coronal domain is indicated

8.3.5 *On the Phase Stability of ACPMs*

In this last results section, the colloidal phase stability of association colloid–polymer mixtures (ACPMs) is assessed in terms of the second osmotic virial coefficient B_2, calculated from the polymer-mediated micelle–micelle interactions. We consider the colloidal particle volume as the one of an isolated micelle and without any added homopolymers, $v_c \equiv v_h^o = (4\pi/3)(R_h^o)^3$. In Fig. 8.9, $B_2^* \equiv B_2/v_c$ is shown for $q = 0.2$ as a function of the homopolymer concentration ϕ_G^{bulk}/ϕ_G^* at fixed homopolymer-to-micelle size $q = 0.2$. These B_2^*-values follow from interactions as those presented in Figs. 8.5, 8.7 and 8.8. We focus first on homopolymers which only interact via excluded volume with the coronal domain, $\chi_{AG} = 0$. Both for crew-cut and starlike micelles, B_2 decreases only weakly upon addition of homopolymer: the depletion zone spans through the coronal domains and hardly on the outside of the micelles, rendering depletion effects weak. The colloidal stability decreases dramatically from $\phi_G^{bulk}/\phi_G^* \approx 0.1$ when depletion effects arise via a corona–homopolymer effective affinity with an enthalpic repulsion beyond the excluded volume ($\chi_{AG} = 0.25$). Hence, it is expected that a suspension of micelles gets unstable at high homopolymer concentrations. Due to the wider coronal region, depletion-induced destabilisation of a starlike micelle–depletant mixture occurs at slightly higher homopolymer bulk concentration. For starlike micelles, the depletion effects are weaker due to the deeper penetration of homopolymer into the corona, in line with our discussion in Sect. 8.3.2.

As hinted at in Fig. 8.7, adding weakly-adsorbing homopolymers ($\chi_{AG} = -0.25$) to a crew-cut micelle suspension leads to a mostly ϕ_G^{bulk}-independent micelle–micelle interaction. For stronger homopolymer–solvophilic block attraction ($\chi_{AG} = -0.5$), bridging attraction and restabilisation is observed in terms of B_2. The trends observed considering (weakly) adsorbing homopolymers show that starlike micelles destabilize more easily than the crew-cut analogues. For adsorbing polymers at high ϕ_G^{bulk}, starlike micelles are more easily destabilized. In micelle applications (such as drug-delivery systems), components with different affinities for the corona blocks may be present. By virtue of their shorter relative hydrophilic block length, we expect that crew-cut micelles are overall more stable in multicomponent systems due to, in essence, a denser and narrower coronal region.

We eventually verified the SCF predictions empirically. The set of interaction parameters used in this study is suitable to describe polycaprolactone-polyethylene glycol (PCL-PEO) diblock copolymer micelles in water [32]. An aqueous suspension of such spherical micelles was prepared ($CL_{12}EO_{45}$, which may be mapped onto our $B_{24}A_{45}$ model crew-cut micelle), and mixed with free PEO homopolymers of different molar masses at different concentrations. A CL block is roughly twice as large as an EO block, which we account for in our SCF comparison. In the left panels of Fig. 8.10 we show photographs of solutions of micelles (diblock copolymer concentration is 5 mg/ml) and pure PEO at fixed concentration for various (weight-

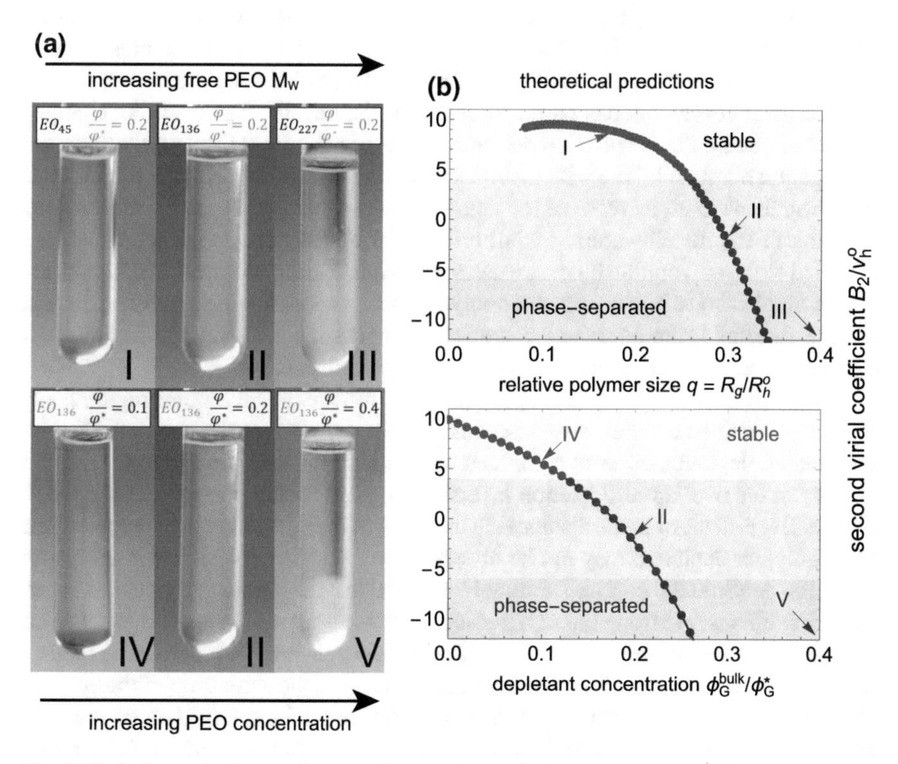

Fig. 8.10 Left panel: pictures shown at the experimental parameters corresponding to the arrows on the right panels; φ is used for the experimental homopolymer concentration. Right panels: phase stability predicted through B_2 from SCF-computations considering $\chi_{AG} = 0$ for the crew-cut micelle studied. **a**: at fixed homopolymer concentration relative to overlap ($\phi_G^{bulk}/\phi_G^* = 0.2$), the relative size of the homopolymer (q) is varied. **b**: at fixed $q = 0.3$, homopolymer concentration is increased

averaged) molar masses: I, $M_w = 2$ kDa; II, $M_w = 6$ kDa; and III, $M_w = 10$ kDa. We use φ to denote experimentally-resolved PEO concentrations, φ^* being the overlap concentration. Photographs IV, II, and V refer to a single type of PEO ($M_w = 6$ kDa, corresponding to $N = 136$) for various PEO concentrations. In the right panels we plot the SCF-predicted B_2 values as a function of the relative polymer size (upper panel) and polymer concentration (lower panel). As experimentally observed, for theoretically-calculated $B_2^* \lesssim -6$ the micellar solution gets unstable. Sample II is stable but clearly more turbid than samples I and IV. This can be explained by a significant free PEO-induced micelle–micelle attraction ($B_2^* \lessapprox 0$), although the attractions are not yet sufficiently strong to induce demixing.

8.4 Conclusions

In this chapter, homopolymer-meditated micelle–micelle interactions were studied. Our approach explicitly considers the associative nature of the micelles. We quantified how the addition of soluble homopolymer in bulk shifts the unimer–micelle equilibrium and studied the effect of different affinities between the homopolymer and the outer coronal blocks on the structure of the micelle. The resulting changes of the bulk unimer concentration determines the micelle properties, particularly the aggregation number. In line with previous experimental observations, homopolymer depletion leads to an increase of the aggregation number which can be rationalised in terms of the micelle–unimer equilibrium. For added homopolymers which are attracted to the solvophilic blocks, weak adsorption leads to a decrease of the aggregation number with increasing homopolymer bulk concentrations. Above a certain affinity threshold, the aggregation number first (dramatically) increases and then decreases with increasing homopolymer bulk concentration.

In case of non-adsorbing homopolymers, the diffuse micellar outer region leads to small effective depletion layers because they penetrate into the coronas. Despite the broader depletion zone as compared to near hard spheres, depletion effects are weaker as the overlap of depletion layers is screened by the presence of the fluffy corona. Even at intermicelle distances in the order of their (undistorted) diameter, full homopolymer depletion may not be observed. Not surprisingly, an added enthalpic repulsion between the corona-forming blocks and the guest homopolymer enhances depletion effects. A strong enough guest polymer affinity to the corona leads to an almost classical polymer-mediated interaction between crew-cut micelles: bridging flocculation at low concentrations, and restabilisation upon saturation of the corona. Remarkably, weakly-adsorbing polymers may not affect the stability of the micellar suspension if the enthalpic and entropic effects of homopolymers in the coronal domain are balanced. For starlike micelles, bridging within the coronal domains leads to attractions which only increase with increasing polymer bulk concentration (up to the near-overlap concentrations studied). The packing of the homopolymer in the corona determines whether the micelle–micelle steric repulsion vanishes at high enough homopolymer concentrations. If the corona thickness is of the order of the added homopolymer diameter, a repulsive contribution to the micelle–micelle interaction is expected to be present.

From the second virial coefficient (B_2) we observe that both crew-cut and starlike micelles are destabilized at high enough non-adsorbing homopolymer concentration due to a (weak) depletion attraction. In case of added adsorbing homopolymer, the B_2-value of the crew-cut micelle–homopolymer mixture is less sensitive than for the starlike case. The trends predicted by our SCF computations for the crew-cut micelles are qualitatively confirmed experimentally using biocompatible PCL-PEO diblock copolymer micelles in water and added free PEO homopolymer. We deduce from our investigation that a narrower coronal domain makes spherical micelles more suitable for applications, such as drug-delivery systems, where multiple macromolecular components are present.

8.A Packing Arguments and Concentration Profiles

The relative volume of a corona block in the micelle per undistorted homopolymer volume (Q) can be derived from geometrical arguments:

$$Q = \frac{g_p R_g^3}{(R_h^o)^3 - (R_A^{max})^3} . \tag{8.7}$$

For $q = 0.2$, we get for the crew-cut micelle $Q = 1.25$ and for the starlike one $Q = 0.3$. We present in Fig. 8.11 the homopolymer concentration profiles upon varying the intermicelle-distance. Homopolymer depletion profiles for isolated ($r \gg 2R_h^o$) micelles were discussed in details in the main text. Homopolymer adsorption, particularly within the corona, is much stronger for the starlike micelle. Both for the crew-cut and starlike micelles, depletion of the homopolymer leads to values of $\phi_G < \phi_G^{bulk}$ at small enough intermicelle distances r. Furthermore, at the same intermicelle distance r depletion effects are clearly stronger for the crew-cut than for the starlike micelle (ϕ_G at $z = 0$ is smaller for the crew-cut micelle for any r). This relates, once again, to the more diffuse peripheral colloidal domain (corona) of the starlike micelle. With increasing ϕ_G^{bulk}, the depletion zones get compressed also when micelles approach each other (except for distances $r < 2R_h^o$). On the bottom panels of Fig. 8.11 homopolymer concentration profiles for adsorption cases are presented. Contrary to the depletion cases, when micelles get close to each other $\phi_G > \phi_G^{bulk}$. While depletion effects are stronger for the crew-cut micelle (near full-depletion for $r = 2R_h^o$), adsorption effects are clearly more pronounced between starlike micelles.

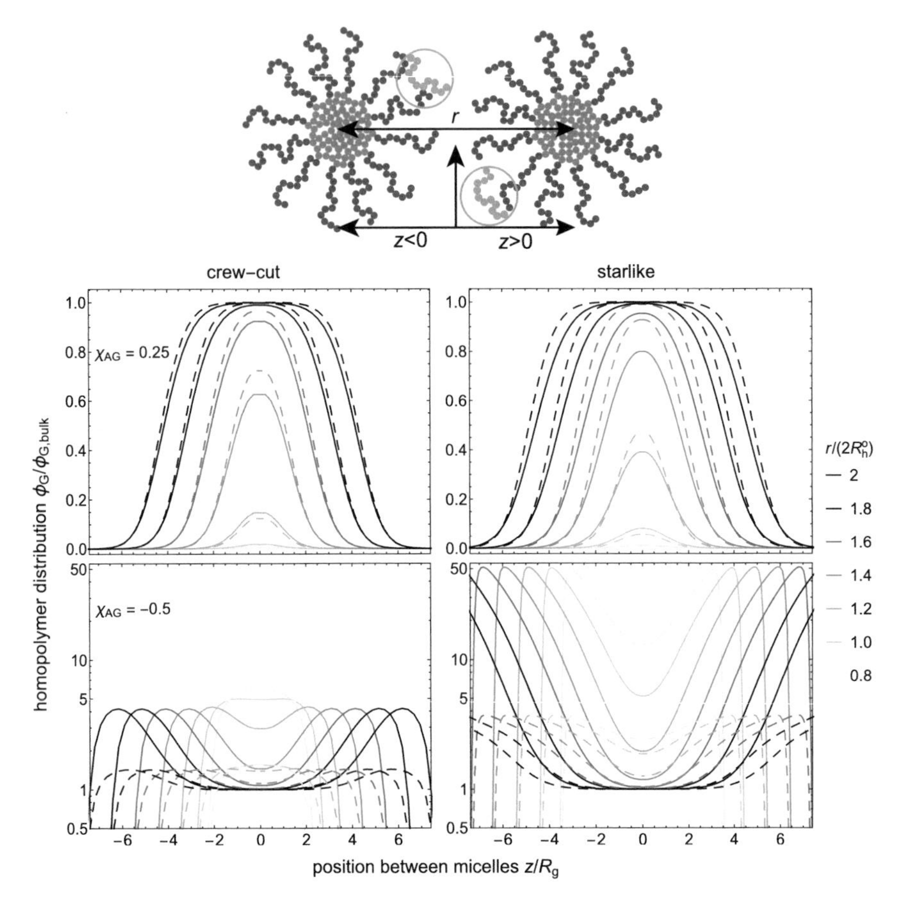

Fig. 8.11 Homopolymer concentration profiles between micelles for depletion (top panels) and adsorption (bottom panels) as a function of the intermicelle distance r as indicated (see sketch on the top, which summarises the distances evoked in the plots). Solid curves correspond to a low homopolymer concentration $\phi_G^{bulk}/\phi_G^* = 0.01$, whereas dashed ones correspond to $\phi_G^{bulk}/\phi_G^* = 0.25$

References

1. F.A.M. Leermakers, J.C. Eriksson, J. Lyklema, in *Soft Colloids, Fundamentals of Interface and Colloid Science*, vol. 5, edited by J. Lyklema (Academic Press, 2005), https://www.sciencedirect.com/bookseries/fundamentals-of-interface-and-colloid-science/vol/5/suppl/C
2. Y. Mai, A. Eisenberg, Chem. Soc. Rev. **41**, 5969 (2012), https://pubs.rsc.org/en/content/articlelanding/2012/cs/c2cs35115c#!divAbstract
3. T. Riley, T. Govender, S. Stolnik, C.D. Xiong, M.C. Garnett, L. Illum, S.S. Davis, Colloids Surf. B: Biointerfaces **16**, 147 (1999), https://www.sciencedirect.com/science/article/pii/S0927776599000661
4. L. Yang, X. Qi, P. Liu, A. El Ghzaoui, S. Li, Int. J. Pharm. **394**, 43 (2010), https://www.sciencedirect.com/science/article/pii/S0378517310003029
5. W. Li, J. Li, J. Gao, B. Li, Y. Xia, Y. Meng, Y. Yu, H. Chen, J. Dai, H. Wang, Y. Guo, Biomaterials **32**, 3832 (2011), https://www.sciencedirect.com/science/article/pii/S0142961211001219

6. J. Wang, X. Xing, X. Fang, C. Zhou, F. Huang, Z. Wu, J. Lou, W. Liang, Philos. Trans. R. Soc. A **371**, 20120309 (2013)
7. D. Lombardo, P. Calandra, D. Barreca, S. Magazù, M.A. Kiselev, Nanomaterials **6**, 125 (2016), https://www.mdpi.com/2079-4991/6/7/125/htm
8. A. Muñoz-Bonilla, S.I. Ali, A. del Campo, M. Fernández-García, A.M. van Herk, J.P.A. Heuts, Macromolecules **44**, 4282–4290 (2011), https://pubs.acs.org/doi/abs/10.1021/ma200626p
9. A. Syrbe, W.J. Bauer, H. Klostermeyer, Int. Dairy J. **8**, 179 (1998). https://doi.org/10.1016/S0958-6946(98)00041-7
10. J. O'Connell, V. Grinberg, C. de Kruif, J. Colloid Interface Sci. **258**, 33 (2003), http://www.sciencedirect.com/science/article/pii/S0021979702000668
11. C.B.E. Guerin, I. Szleifer, Langmuir **15**, 7901 (1999)
12. I. Hamley, *Block Copolymers in Solution: Fundamentals and Applications* (Wiley, New York, 2005)
13. R. Savić, T. Azzam, A. Eisenberg, D. Maysinger, Langmuir **22**, 3570 (2006), https://pubs.acs.org/doi/10.1021/la0531998
14. G.J. Fleer, M.A. Cohen Stuart, J.M.H.M. Scheutjens, T. Cosgrove, B. Vincent, *Polymers at Interfaces* (Springer, Netherlands, 1998), pp. XX, 496
15. B. Vincent, J. Edwards, S. Emmett, A. Jones, Colloids Surf. **18**, 261 (1986), http://www.sciencedirect.com/science/article/pii/0166662286803171
16. S. Abbas, T.P. Lodge, Phys. Rev. Lett. **99** (2007), https://journals.aps.org/prl/abstract/10.1103/PhysRevLett.99.137802
17. R. Yamazaki, N. Numasawa, T. Nose, Polymer **45**, 6227 (2004), https://www.sciencedirect.com/science/article/pii/S0032386104004872?via%3Dihub
18. S. Abbas, T.P. Lodge, Macromolecules **41**, 8895 (2008)
19. Y. Lauw, F.A.M. Leermakers, M.A. Cohen Stuart, O.V. Borisov, E.B. Zhulina, Macromolecules **39**, 3628 (2006), https://pubs.acs.org/doi/abs/10.1021/ma060163t
20. E.B. Zhulina, O.V. Borisov, Macromolecules **45**, 4429 (2012). https://doi.org/10.1021/ma300195n
21. M.L. Kurnaz, J.V. Maher, Phys. Rev. E **55**, 572 (1997), https://journals.aps.org/pre/abstract/10.1103/PhysRevE.55.572
22. A. Quigley, D. Williams, Eur. J. Pharm. Biopharm. **96**, 282 (2015), https://www.sciencedirect.com/science/article/pii/S0939641115003288
23. A. Mulero, *Theory and simulations of Hard-Sphere Fluids and Related Sytem* (Springer, Heidelberg, 2008)
24. G.A. Vliegenthart, H.N.W. Lekkerkerker, J. Chem. Phys. **112**, 5364 (2000), https://aip.scitation.org/doi/10.1063/1.481106
25. Č. Koňák, Z. Tuzar, P. Štěpánek, B. Sedláček, P. Kratochvíl, in *Frontiers in Polymer Science*, edited by W. Wilke (Steinkopff, Darmstadt, 1985), pp. 15–19, https://link.springer.com/chapter/10.1007/BFb0114009
26. M. Villacampa, E. Diaz de Apodaca, J.R. Quintana, I. Katime, Macromolecules **28**, 4144 (1995), https://pubs.acs.org/doi/abs/10.1021/ma00116a014#citing
27. T. Yoshimura, K. Esumi, J. Colloid Interface Sci. **276**, 450 (2004). https://doi.org/10.1016/j.jcis.2004.03.069
28. W. Li, M. Nakayama, J. Akimoto, T. Okano, Polymer **52**, 3783 (2011). https://doi.org/10.1016/j.polymer.2011.06.026
29. T. Zinn, L. Willner, R. Lund, V. Pipich, M.-S. Appavou, D. Richter, Soft Matter **10**, 5212 (2014). https://doi.org/10.1039/C4SM00625A
30. K. Mortensen, Europhys. Lett. **19**, 599 (1992). https://doi.org/10.1209/0295-5075/19/7/006
31. K. Mortensen, J. Phys.: Condens. Matter **8**, A103 (1996). https://doi.org/10.1088/0953-8984/8/25a/008
32. A. Ianiro, J. Patterson, Á. González García, M.M.J. van Rijt, M.M.R.M. Hendrix, N.A.J.M. Sommerdijk, I.K. Voets, A.C.C. Esteves, R. Tuinier, J. Polym. Sci., Part B: Polym. Phys. **56**, 330 (2018), https://onlinelibrary.wiley.com/doi/full/10.1002/polb.24545

33. H.N.W. Lekkerkerker, R. Tuinier, *Colloids and the Depletion Interaction* (Springer, Heidelberg, 2011)
34. C. Guerrero-Sanchez, D. Wouters, C.-A. Fustin, J.-F. Gohy, B.G.G. Lohmeijer, U.S. Schubert, Macromolecules **38**, 10185 (2005), https://pubs.acs.org/doi/10.1021/ma051544u
35. R.Tuinier, S. Ouhajji, P. Linse, Eur. Phys. J. E **39** (2016), https://link.springer.com/article/10.1140/epje/i2016-16115-5
36. A. Jones, B. Vincent, Colloids Surfaces **42**, 113 (1989), http://www.sciencedirect.com/science/article/pii/0166662289800812
37. P. de Gennes, *Scaling Concepts in Polymer Physics* (Cornell University Press, Ithaca, 1979), https://books.google.nl/books?id=Gh1TcAAACAAJ
38. G.J. Fleer, A.M. Skvortsov, R. Tuinier, Macromolecules **36**, 7857 (2003), https://pubs.acs.org/doi/abs/10.1021/ma0345145
39. R. Tuinier, G.A. Vliegenthart, H.N.W. Lekkerkerker, J. Chem. Phys. **113**, 10768 (2000). https://doi.org/10.1063/1.1323977
40. A. Ianiro, Á. González García, S. Wijker, J.P. Patterson, A.C.C. Esteves, R. Tuinier, submitted. https://pubs.acs.org/doi/10.1021/acs.langmuir.9b00180

About the Author

Yo soy yo y mi circunstancia
y si no la salvo a ella no me salvo yo.

I am I and my circumstance;
and, if I do not save it, I do not save myself.

José Ortega y Gasset

I was born in Los Silos, northwestern Tenerife, in 1990. There, I attended the local public primary school and high-school. I did a bachelor in Physics in Universidad de La Laguna (Tenerife), with a one-year exchange programme with Universidad de Granada (Andalucía) (2008–2012). In Granada, I did a short bachelor end project on the viability of carbon nanotubes as drug-delivery systems, supervised by Prof. J. Lopez-Viota Gallardo and Prof. Á. Delgado Mora. Following the recommendation of my supervisors in Granada, I moved to Utrecht University where I did a master in Nanomaterials Science (2013–2015). I conducted my master project research, a simulation account on random packings of anisotropic particles, at the Van't Hoff Laboratory under the supervision of dr. R. Baars and Prof. A. P. Philipse. As part of this master, I did an internship at DSM-Geleen under the guidance of dr. L. Bremer, where we worked on an alternative method for characterising soft matter. After a short break, I started my Ph.D. studies under the instruction of Prof. Dr. Ir. R. Tuinier in between the Van't Hoff Laboratory (Utrecht) and the newly-formed Laboratory of Physical Chemistry (Eindhoven). Part of the results of these Ph.D. studies are presented in the thesis, and have been published in peer-reviewed journals and presented in Dutch and international conferences.

© Springer Nature Switzerland AG 2019

Á. González García, *Polymer-Mediated Phase Stability of Colloids*,
Springer Theses, https://doi.org/10.1007/978-3-030-33683-7

Printed in the United States
By Bookmasters